An Hachette UK Company
www.hachette.co.uk

First published in Great Britain in 2015 by Hamlyn, a division
of Octopus Publishing Group Ltd
Endeavour House
189 Shaftesbury Avenue
London
WC2H 8JY
www.octopusbooks.co.uk

Conceived, designed and produced by
Quid Publishing
Level 4, Sheridan House,
114 Western Road
Hove BN3 1DD
www.quidpublishing.com

Designed by Clare Barber
Written by Chris Graham

ISBN 978-0-600-63013-5

A CIP catalogue record for this book is available from the British Library

Printed and bound in China

10 9 8 7 6 5 4 3 2 1

Every effort has been made to ensure that all of the information
in this book is correct at the time of publication.

THE
CHICKEN
KEEPER'S

PROBLEM SOLVER

100 COMMON PROBLEMS EXPLORED AND EXPLAINED

CHRIS GRAHAM

hamlyn

CONTENTS

INTRODUCTION

Although this book concentrates on problems, don't imagine for one minute that this is all you are going to encounter with your chickens. Quite the contrary, in fact; for most of the time, keeping a few hens in the back garden is as trouble-free as it is delightful.

However, things can and do go wrong from time to time, and when trouble strikes, it is important that you know how best to deal with it. Taking the appropriate steps in a decisive and confident way is the name of the game. Good chicken husbandry is all about staying one step ahead; recognising problems at the earliest possible opportunity, then doing something useful before things get too serious. The key to this, of course, is learning to appreciate what might go wrong, and understanding the tell-tale signs when it does happen.

While most chickens are a good deal more resilient than their feathery and often delicate outward appearance may suggest, they are all intolerant to stress. Hens that feel stressed will be unsettled and reluctant to lay regularly. They will also be significantly more susceptible to infection, disease and the debilitating effects of any underlying health issues that their bodies may be harbouring. The trouble is that stress can be caused by any number of triggers, from things as apparently inconsequential as a change in feed type or run location, to more serious issues such as bullying or overcrowding.

The purpose of this book is to walk you through 100 of the most common concerns likely to affect your garden hens and to provide practical solutions for dealing with them. You can dip in and out, referring to specific issues as and when you need to, or simply read and learn from the complete collection in one go. Either way, I trust that the knowledge you gain will enhance your own chicken-keeping experience as well as bolster the health and welfare of your hens.

It's essential to appreciate that hens do all they can to disguise the effects of illness or injury; like any other prey species, they are desperate not to appear outwardly weak. However, this 'deception' can make it that much harder for inexperienced keepers to spot when things are going wrong. The secret is to develop a decent working knowledge of chicken behaviour, and the practical problems and threats from disease and parasite activity that they can face. Cover these basics and you'll be well on your way to a hassle-free and enjoyable time with your back-garden chickens.

Much of what you will need to learn comes down to good old-fashioned common sense. Put yourself your birds' place and ask yourself whether or not you'd be happy with their living conditions. If, like me, you would hate having to paddle around barefoot in muddy puddles all day, scratching on bare, featureless earth or having to share a stuffy, cramped and dirty house with too many others, then don't expect your birds to put up with such conditions either.

While there's certainly no substitute for practical experience, everyone has to start somewhere, which is where books like this one can prove so useful. Getting the basics clear in your mind as early as you can is bound to help you make the right choices when you need to. Taking the correct action at the proper time is the mark of a good chicken keeper and, most of the time, this involves being pro-active rather than reactive. Problems with chickens rarely get better by themselves; appropriate action will need to be taken sooner rather than later and you, as the birds' responsible keeper, are the one who needs to take it.

CHAPTER ONE

FOOD AND WATER

The two factors that will most govern the success you enjoy with your back garden chickens are the quality of the feed you give the birds and your ability to maintain a constant supply of fresh and clean drinking water. Plenty of new keepers fail to appreciate the importance of getting these two aspects right, as well as the fact that a regular water intake is actually more important to a chicken's well-being than food. Their tolerance to dehydration is far lower than it is to a lack of food; a chicken can lose virtually all its body fat and still recover, whereas if it loses any more than 10 per cent of its water content, it will prove fatal. The provision of food and drink must therefore be at the very heart of your good husbandry routine. Both must be checked daily and maintained to the highest standards to ensure that your chickens enjoy the best quality of life.

Consistency is key – any sudden change could have a detrimental effect on their appetite. Likewise, appetite is a useful barometer for your chickens' state of health; if any seem to be off their food it is a cause for concern, and warrants further investigation. Then there are the daily challenges of making sure supplies of food and water are readily available, that feeders and drinkers are kept clean and free from bacterial infection or mould, and that wild birds or rodents aren't helping themselves to a free meal.

01 My chickens have gone off their regular feed

CAUSE

The feeding of 'treats' and the seasonal variation in the availability of insect life for free-ranging birds are both factors that will affect their appetite for standard layers pellets.

SOLUTION

There is an unfortunate tendency among back-garden hen keepers, especially those who are still new to the hobby, to feed their birds treats. The high levels of both sugar and salt in processed food intended for human consumption is completely at odds with a chicken's dietary requirements. Chickens should not be fed this sort of material; it can actually be harmful because their digestive systems simply aren't designed to cope with it. Nevertheless, the practice goes on, very often to excess. The birds fill up on these inappropriate foodstuffs and, in comparison, the standard layers' ration becomes very plain and unappetising. Quite apart from the health risks associated with eating these inappropriate 'treats', birds that stop eating their formulated feed can develop vitamin and mineral deficiencies that will adversely affect egg quality, laying performance and overall health.

There's also a tendency for free-ranging hens to find more of their own food (in the form of insects and grubs, and so on) during the spring and summer months, and to naturally eat a little less of their regular ration. But this, of course, is to be expected and is perfectly acceptable; these 'natural treats' are full of nourishment.

Chickens thrive on routine, so the solution is to be consistent with your feeding. Don't chop and change what you give them, resist the temptation to feed unsuitable treats and make sure the formulated ration you provide is of good quality and fresh.

Avoid the temptation to feed your birds 'treats', even if you get the impression that they love eating them. Restrict supplementary feeding to a handful of mixed corn or fresh fruit and vegetables every now and then.

BE SMART WITH FEED

Many beginners fail to appreciate the importance of matching the chicken feed they provide to the type of birds being fed. 'Layers', 'growers' and 'fatteners' will all benefit from an appropriately formulated and nutritionally balanced feed.

02 Feed is going mouldy in the sack

CAUSE

Modern formulated chicken feeds have a shelf life. Once they pass their 'best before' date the oils and fats they contain can turn rancid, which may lead to the formation of potentially harmful mould.

SOLUTION

It is important to be aware of the 'best before' date for the chicken feed you are using, and to ensure that it is all consumed in good time to avoid wastage. This obviously requires an understanding of typical consumption rates, and a sensible approach when ordering. While it is always cheaper to buy chicken feed in bulk, the value of any savings made will be lost completely if the product goes out of date before you've had a chance to use it. Chickens simply won't want to eat feed that's gone off, and mouldy feed can be seriously damaging to their health if consumed. Feed storage is another important factor in this respect. Sacks must be kept in dry, temperate surroundings, and always out of direct sunlight.

Take care with your feed and make sure that you store it in a garage or shed that does not get too hot and is dry. The feed you give to your hens represents the most significant, ongoing cost associated with the hobby; as prices continue to rise, the last thing you want to do is to throw feed away for no good reason.

There really is no excuse for letting chicken feed go mouldy inside its sack. Not only is it a terrible waste of money, it can make your hens ill, too.

03 Wild birds are eating the chickens' food

CAUSE

Large wild birds, like crows and rooks, really can make a dent in your chicken feed reserves if they are easily accessed.

SOLUTION

With chicken feed prices rising ever higher, the last thing you want is to extend an open invitation for a free lunch to the local wildlife, especially the winged ones. Traditional feeders are just as easy for wild birds to use as they are for your chickens, so there is nothing to prevent them tucking in.

Hanging up old CDs to flash in the wind can ward off unwanted avian diners. Raising the feeder off the ground can also help, so that the feed itself remains unreachable for even the larger wild birds, but is still comfortably within range of your hens. Alternatively, look at some of the more ingenious feeder designs available nowadays. There are wall- or fence-mounted options that are well protected against thievery from perching birds; these can be mounted high enough to protect their content from ground-based raids, too.

You can also buy clever treadle-operated feeders that store the supply of feed securely under a cover until the weight of a chicken standing on the treadle swings the lid open. The bird can feed to its heart's content; then, as soon as it steps off the treadle, the lid closes again and the feeder is secured once more.

A treadle-operated feeder like this one is great for minimising food theft by wild birds and rodents. The chickens will need to be trained to use it, though.

04 Too much feed is being spilt

CAUSE

Using the wrong type of feeder can contribute significantly to the amount of expensive poultry feed that gets wasted.

SOLUTION

Spilled chicken feed is not only a waste of money, it also attracts wild birds, rats and mice – and these unwelcome visitors can bring parasites and disease with them. So, quite apart from taking extra care when refilling feeders to avoid accidental spillages, it is important that you equip yourself with a decent feeder.

There are plenty of feeder options on the market nowadays, but not all of them are particularly effective at helping to control wastage. Straightforward, open-topped troughs and bowls are probably the worst, because the sideways flicking action that many hens adopt when they peck tends to result in pellets and mash being thrown out on to the surrounding ground. Most of the gravity-fed types feature anti-waste grilles around the feeding channels, which work to compartmentalise small sections of feed to minimise the effects of sideways flicking.

Another useful addition can be a feeder stand featuring a mesh top, which allows any spilt feed to fall down out of harm's way so that you can clear it up later.

05 New feed is not being eaten

CAUSE
Sudden changes to the feeding regime are likely to upset the eating patterns of laying hens.

SOLUTION
Chickens are creatures of habit that thrive on routine; this aspect of their character should impact on your good husbandry routines in many different ways. Changes to the birds' usual feeding regime can come as something of a shock to them. In fact, this can be a significant enough disruption to stop them eating altogether.

Whether you've decided to switch between chicken feed brands for reasons of cost, because you've found a more convenient supplier, because there is a new product on the market that you fancy trying or because you're switching from one age group to another, you should never simply introduce the new feed in one go. You need to impose a gradual transition over a period of a week or so.

As a guide, begin with an 80:20 mix of the existing feed with the new product; then, once this has been consumed, adjust to a 60:40 split. A day or two later, mix up a new batch at a 40:60 ratio, in favour of the new feed, followed by a 20:80. Once this has been successfully consumed, your birds will be ready to receive a full dose of the replacement product, and you will have successfully made the change with the minimum disruption.

06 One bird is not eating

CAUSE

Individual hens that stop eating may be being bullied, or they may have a physical ailment such as crop impaction, sour crop or gapeworm.

SOLUTION

One of the reasons why it is very important to spend a few minutes every day simply observing your chickens is so that you can make sure they are all behaving normally. A good time to do this is first thing in the morning, when the hen house is opened up for the new day. The birds should emerge enthusiastically and with reasonable vigour, their tails held high. Most should make a beeline for the feeder, and any that don't deserve a little special attention.

Birds that obviously keep themselves to themselves and that are reluctant to gather around the feeder with the rest may be being bullied, so check them over for any signs of feather damage or pecking wounds to the comb or wattles. Birds with injuries that have drawn blood should be removed from the pen immediately, treated where necessary and allowed time for the wounds to heal before being reintroduced.

Also, try to check the condition of the droppings of birds that are not eating, and look – or better still, feel – for hard or soft swellings towards the base of the neck. Abnormal dropping consistency is always an indicator of internal trouble, while lower neck swelling can point towards sour crop (soft swelling) or crop impaction (hard swelling). Both will inhibit appetite and require veterinary attention.

KEEP AN EYE ON THEM

Paying regular, close attention to your birds is at the heart of good chicken keeping. Not only will it allow you to keep an eye on their condition, it will also help your hens to feel comfortable and relaxed in your presence; both are vital factors contributing to a healthy, settled flock.

Feeding pellets from the hand can be a great way of building confidence among your hens.

07 The feeder is always empty

CAUSE

Either you're simply not providing enough food for your chickens, or the wild birds and animals are helping themselves.

SOLUTION

Food management is an important issue for back-garden chicken keepers. The risks associated with overfeeding your birds are very real and potentially serious, especially among birds that get minimal exercise because they're contained in a small house/run combination, for example. The generally accepted feeding ration is 125g per bird, per day, but most keepers don't bother to weigh this out and simply feed their birds on an ad hoc basis, replenishing the feeder as and when necessary.

If your birds are regularly exhausting your feeder, then either you're not providing enough food, or you've got a major issue with wild birds and/or rodents stealing from the feeder. You have a legal obligation to keep chickens adequately fed so, if you're in any doubt about the amount of feed being provided in relation to the number of birds being kept, simply add another feeder. A bit of careful observation should tell you whether the local wildlife is causing a problem. Birds are pretty easy to spot, but you'll need to be a good deal more patient to see rats or mice in the act of thieving. Droppings near the feeder can be a useful sign.

If you find rodents to be a particular problem, then one sure-fire way to combat such unwelcome losses is to remove the feeder from the chicken run at night. With your hens shut securely inside their house, they certainly won't have any need for it, and there is no point in leaving it at the mercy of greedy, nocturnal visitors. Moving it into a garden shed or utility room should eliminate this sort of costly wastage.

Match the feeder size you are using to the number of birds you have to ensure the food is available ad hoc.

08 The hens' bedding is wet

CAUSE

Wet bedding inside the hen house may be caused by the presence of drinkers, a leaking house or bad weather.

SOLUTION

There is a common misconception that chickens benefit from the provision of food and water inside the hen house during the night. Both are completely unnecessary as chickens will neither feed nor drink overnight. Drinkers inside the house are bad news, full stop. Anything that increases the chances of wet bedding material must be avoided; hens using a drinker left permanently inside their house are bound to cause a spill at some point.

Leaking roofs, poorly-fitting or warped doors and split side panels all provide opportunities for rainwater to get inside the house; it will typically run down the inside of the walls to settle on the floor and start soaking the bedding from the bottom up. Look carefully for suspicious staining and don't delay when repairs are necessary.

Finally, moisture and wet mud and other debris will be carried into the house by the birds during bad weather. If the run is in poor condition, with mud and puddles, the situation will obviously be that much worse. The only answer is to clear out and replace the bedding as often as is necessary to keep it dry during wet spells of weather. A layer of gravel, wood chip or even fresh straw at the bottom of the pop-hole ramp can help to clean the birds' feet as they approach the house.

HEN HOUSE BASICS

The hen house you are using does not have to be a fancy affair: your hens will not mind if it cost £150 or £1,500. As long as they have enough space, nest boxes to use and are kept secure at night, dry and draught-free, they will be happy.

Wet bedding inside the hen house is a real no-no and should be avoided at all costs. Any structural damage that allows water to get inside must be repaired without delay.

09 The drinker is turning green

CAUSE

If drinking water isn't changed regularly enough – especially during mild or warm weather – green algae and bacterial growth can form inside the drinker and in the drinking lip.

SOLUTION

Dirty, stained drinkers are always a sign of poor husbandry standards, and can pose a threat to the health of the birds using the drinker due to bacterial infection. There really is no excuse for not keeping these essential pieces of equipment clean and fresh, like the water that should always be inside them.

Ideally, chicken drinkers should be thoroughly cleaned on a weekly basis, using a recognised but non-toxic cleaner and disinfectant. Many chicken keepers recommend using products that are used to clean and sterilise babies' milk bottles, which are gentle yet effective.

If things get really bad, and you find yourself having to use a heavy-duty cleaner or bleach-type product, then make sure you thoroughly rinse out the drinker several times afterwards, before refilling it with fresh water for the birds.

NATURAL CLEANER

Apple cider vinegar is a traditional, natural product that acts as a great digestive tonic for chickens. Adding a teaspoonful or two to their water can help to keep the inside of the drinker clean, too. However, this is only applicable to plastic drinkers; it should never be added to the water inside galvanised metal drinkers as it will react with the coating on the metal.

Drinkers must be kept clean and filled with fresh water. Mould and bacterial growth will taint the water and make your birds ill.

10 Ice has caused the drinker to crack

CAUSE

This is a common problem on cold winter nights, when water that's left inside a plastic poultry drinker freezes and turns into ice. As the water freezes it expands, cracking the usually thin plastic of these vessels with ease and rendering the drinker useless.

SOLUTION

You have two options to avoid this annoying event. Simplest of all, when using the common, plastic drinkers, is to take the trouble to empty them every night when you think there's a risk of frost. In fact, the benefits of doing this are threefold. For a start, it will stop the drinker being damaged by the formation of ice inside. Secondly, it will ensure that your birds enjoy a good supply of fresh, clean water the following day when you have to refill the drinker. Thirdly, there are few things more irritating than having to mess about trying to defrost a frozen drinker when time is short in the morning, and you're rushing to get off to work or run the children to school; simply filling the drinker with fresh water is much quicker.

Alternatively, switch to using metal drinkers. These, by their very nature, are much more resistant to cracking, but will still need to be defrosted in the morning if they haven't been emptied the previous night. They are more expensive to buy as well but, by and large, do tend to last a lifetime. Plastic drinkers, on the other hand, tend to have a useful life of no more than four or five years, although this varies according to their quality and how they are treated. So it really makes sense, whatever sort of drinker you use, to empty it out (or bring it inside) in icy weather.

The cheapest plastic drinkers are made from very thin material that will turn brittle quickly. They will also be most prone to cracking if left full in icy weather. Sturdy, well-made units like this one can withstand extremes in weather.

FALSE ECONOMIES

With the internet awash with chicken-keeping equipment, it can be hard not to start buying on price. If you can possibly avoid it, though, resist the urge to seek out the lowest price. Never forget that feeders, drinkers and hen houses at the bargain basement end of the market cost next to nothing for a reason...

11 I caught my chickens eating a mouse

CAUSE

Chickens are omnivores, so they are happy to consume both vegetation and animal protein.

SOLUTION

It may not be something that many keepers like to see but, given the opportunity, their cute, feathery and otherwise friendly hens will be more than happy to devour a worm, frog, dead mouse or even a wild songbird. The only way to stop your chickens from doing this is to contain them in an enclosed run which, unless very carefully managed, can throw up health and welfare issues for the birds.

Most chickens are great foragers, and love nothing more than getting outside to scratch and peck at anything they can find in nature's larder. In fact, they derive significant benefits from this sort of behaviour. They glean nutritious dietary supplements by consuming insect and other animal proteins they may find, and the activity involved in foraging is great for muscle development and general health and well-being.

The mental effects are also very important. Chickens are natural free-rangers. In the wild, their ancestors spent the majority of their time finding their own food; it came naturally to them and was a basic survival instinct. This preference remains a deep-seated desire in domestic chickens to this day, so depriving them of the opportunity to do so can lead to significant problems. Hens that are unable to get out and about become bored, which increases stress levels and makes them more likely to turn on each other. Also, stressed birds are far more likely to develop health problems, as their normal resistance to disease and infection will be lowered. So the answer to this problem is – just leave them to it!

Don't be surprised to find your hens feasting on all sorts of tasty morsels from Mother Nature's larder.

AS RARE AS...

The old adage about something being 'as rare as hens' teeth' is, of course, based on the fact that hens don't have any teeth at all. Instead, all the food that they swallow is broken down by the grinding action of previously swallowed grit contained within a muscular organ called the gizzard.

12 I'm worried about my free-range hens eating poisonous plants

CAUSE

There is a lot of speculation on the internet about which plants chickens can and cannot eat safely and, as is the way of these things, plenty of it is inaccurate.

SOLUTION

Like most wild creatures, domestic chickens seem to have an innate sense about what they can and cannot safely eat. I have never come across a single case of a chicken having been poisoned from eating something it should not have, even though there are many plants commonly found in gardens that are perfectly capable of doing harm.

Daffodils are one of many plants classed as poisonous to chickens. We have a few clumps of daffodils in our chicken enclosure, which flower beautifully each spring without any drama. On odd occasions I have seen signs that the hens have pecked off some of the bright yellow flower heads, but that is as far as it goes. I have certainly never been tempted to removed the daffodils from the area and, given that the birds never make contact with the actual plant bulbs, I do not consider their presence to be a problem.

However, as a matter of interest, here is a list of some of the commonly found garden plants that are considered to be potentially harmful to chickens: bracken, buttercup, daffodil, delphinium, foxglove, horse chestnut, hyacinth, hydrangea, ivy, lupin, oleander, rhododendron, rhubarb, tulip, wisteria and yew.

Chickens that are allowed to range freely in the garden will encounter all manner of plants that are potentially harmful. Fortunately, though, they generally seem to have the sense to avoid eating things they shouldn't.

CHAPTER TWO

HOUSING

With so many options available on the market these days, buying the right poultry house can be a confusing business, but it is essential to get this decision right. The unit must be the right size for the number of birds you plan on keeping in it, and it needs to provide them with secure, dry and draught-free accommodation. Ideally, it should also be large enough that the birds are able to spend time comfortably inside it during the day when the weather is bad.

Bird capacity is commonly calculated by available perch length, with manufacturers and sellers typically working on the basis of 20cm per bird. This is all well and good as far as roosting is concerned, but pays no heed to how suitable the house may be for occupation during the day. In most cases, actual floor area in modern, back garden-type hen houses is minimal because they're designed as sleeping quarters only. This is definitely something to bear in mind when buying, especially if you live in an area with a challenging climate.

Once you have chosen your hen house, you need to pick a good location for it, and make sure that it is maintained to the highest possible standard. Routine regular checks are needed to look out for any leaks or other damage, and any necessary repairs should be carried out promptly. Your hens are much more likely to flourish if they are provided with a secure and happy home.

13 An external nest box is leaking

CAUSE

Many wooden hen houses with external nest boxes feature poorly designed hinged lids with over-sized gaps and opening clearances that allow in both wind and rain.

SOLUTION

If water is getting into your nest box (or anywhere else in the hen house, for that matter), then remedial action needs to be swift and effective. The first job is to identify how and why the water is getting in. In most cases, this will be related to the lid. It might be that the screws used to fix the hinges have gone right through the thickness of wood, allowing water in via the screw holes.

More usually, though, the problem will be one of fit, or lack of it. Gaps typically exist where the back of the lid meets the wall of the house. Depending on the type of hinges used, varying degrees of clearance are required here to allow the lid to be lifted up fully and to ensure easy access for collecting eggs. It only takes a slight misalignment with the hinges, or for the wood itself to warp, to start letting the water in.

In this instance, one of the easiest and simplest remedies is to tack a suitably sized piece of canvas or rubberised sheet to the wall of the house, above the nest box roof, and let it hang down to bridge any gap so that rain water gets channelled over the entry point to drain straight off the edge of the nest box roof.

Tacking on an off-cut of pond liner material is the simplest solution for a leaking nest box roof.

OUTWARD APPEARANCES

Just because a hen house looks good from the outside does not mean that it's fit for purpose on the inside. These units represent a major investment, so buyers need to be sure that what they purchase is practical and well thought out, rather than simply an attractive but inadequate imitation of some other design.

14 The house bedding smells unpleasant

CAUSE

House bedding only takes on a bad odour if it is left for too long before it is changed and the house is cleaned out.

SOLUTION

Chickens shouldn't smell unpleasant, and neither should their run or house. The only chickens that do are the poor creatures that are badly looked after by keepers who either don't care enough, or who fail to appreciate the importance of a good husbandry routine. 'Little and often' really is one of the key rules that applies to looking after chickens at home. It needn't be time-consuming if you're vigilant and can set aside a few minutes each day to keep on top of the basics.

'Poo-picking' inside the hen house on a daily basis as you collect the eggs takes only a matter of seconds, but really does make a big difference to the quality of life your birds will enjoy. The objective should always be to keep the in-house environment as fresh as possible, so that the time your birds spend inside is both healthy and relaxing. Using a pair of rubber or gardening gloves – or a small hand scoop – to remove the droppings clumps into a compost bucket, is easy, quick and very worthwhile.

Chickens can suffer greatly from stress; a poor environment is a big contributor to this. Allowing the house bedding to become badly contaminated with droppings will promote the formation of ammonia, which has a bad odour and can cause all sorts of problems for the birds, including eye and respiratory complications, as well as burns to the skin. As dirt levels rise, so will the chances of bacterial growth and the development of mould, another source of nasty aromas and health risks for the birds.

LET IT ROT

Chicken droppings are great for making compost; add a 5cm thick layer to every 15cm of other matter, and let it rot for the best results. The droppings are relatively alkaline, so while they can work well on plum trees, blackcurrants and vegetables, they're not suitable for use on camellias, azaleas, rhododendrons and other lime-loving plants.

It is essential that the bedding in your hen house is clean and fresh at all times. If it is dirty enough to put you off sitting in the house, then why should your hens have to endure it?

It's really dusty inside the hen house

CAUSE

Dusty environments inside the hen house result from the use of the wrong sort of floor bedding, and can be intensified by inadequate and/or infrequent cleaning.

SOLUTION

Maintaining a healthy environment inside the hen house is one of the key principles of good chicken husbandry. It is a requirement that's all too often overlooked by beginners who wrongly assume that all they need to do is to provide a house and throw in some bedding from time to time. But keeping things up to scratch inside is something of a balancing act. Moisture and humidity must be kept in check because of the very real threat they pose to the birds' health (see problem 18). Likewise, dust can cause serious problems; it can aggravate the birds' eyes and lead to respiratory difficulties. This is one of the reasons why it is not advisable to use hay as a bedding material or, to a lesser extent, straw. Both are very dry materials that can be a source of harmful dust and spores. Sawdust is another option that some keepers favour, despite the fact that it can contain a significant amount of dust.

Nowadays, though, it is possible to buy a range of materials that have been treated specifically to remove any dust content, and these really are the best option for poultry house bedding (both on the main house floor, and for use in the nest boxes). Some are even scented, or treated with non-toxic disinfectants or anti-viral and anti-fungal agents, which can add valuable extra performance to the bedding material – as well as being super-absorbent, they help to maintain a dry and fresh environment and actively work to fight against the formation of bacteria and moulds. There is obviously a premium to be paid, but many keepers now consider that the extra peace of mind they offer justifies the expense.

*House bedding –
whichever type you choose
– needs to be a dust-free
product to help ensure the
best in-house environment
for your hens.*

TAKE CARE

Although seemingly innocuous and, in most cases, without
much appreciable smell, chicken droppings do have the
potential to cause illness if handled carelessly. Always take
sensible precautions when handling old house bedding; wearing
gloves is the best approach but, if you don't, then always wash
your hands thoroughly afterwards before doing anything else.

16 The house door won't close

CAUSE

Hinges can become clogged with damp bedding and droppings or, more seriously, the wood of the door can warp, affecting the alignment.

SOLUTION

The safety of your hens at night can be completely dependent on the security of their hen house. The rigidity of its construction, the quality of the materials used and the way in which it has been put together will all play a part. Pop holes and doors need to be lockable with tamper-proof bolts; wily foxes have been known to open inadequately secured hen house doors. The integrity of the whole structure is a vital issue.

While this won't usually be a problem with well-specified houses from established manufacturers, it may not be the case with some of the cheaper, flat-packed options that are flooding the market nowadays. Much of the problem stems from the poor quality of the wood used to make them – it's often too thin, poorly seasoned and badly treated. These shortfalls can result in warping and splitting at the first hint of bad weather. Expose some of these units to the rigours of a wet winter and alignments become distorted, hinges and fixings bend and gaps appear.

There will usually be some scope for careful readjustment to deal with these issues, and splits can be covered to seal them against the weather. However, the reality is that a house badly affected in this way simply isn't going to last. It may often be found that the door furniture needs to be replaced with better-made, corrosion-resistant alternatives. The use of thin timber can make effective fixing difficult, too (thin, spindly and easily broken screws are often the preferred choice among house manufacturers that are intent on keeping production costs to a minimum). Additional timber battening attached from behind may often be necessary to allow the fixing of better hinges, held in place with larger screws.

DEFENSIVE MEASURES

In a well-organised poultry set-up, the security you have on the hen house really should be regarded as a second line of defence. Your primary objective should always be to keep predators out of the chicken enclosure, so that they never get a chance to test the security of the chicken coop.

To keep your hens safe, ensure that faulty or broken locks, split wood or breakages are repaired as soon as possible.

17 The roofing felt has split

CAUSE

This tends to be an age-related problem, although deterioration can be accelerated by interference from the chickens and wild birds.

SOLUTION

Given the importance of the roof on your hen house, it is vital that it remains 100 per cent functional at all times. While a number of house manufacturers now employ specialist roofing materials such as Onduline (typically offering a longer-lasting and less mite-friendly alternative), many still opt for the traditional approach: a wooden panel covered in roofing felt. This material has been used for centuries and is essentially bitumen-covered fabric, with the addition of limestone, sand or modern polymers to strengthen the coating. It provides a completely waterproof covering and works well for the first few years.

However, as time passes, the coating hardens and starts to become brittle. The regular heating and cooling caused by the summer sun and winter frosts promotes puckering which, in turn, applies stresses that can trigger surface cracking. Problems can be accelerated by the scratching of chickens that flutter up on to the roof, and the pecking of wild birds. It is essential that you keep a close eye on the condition of roofing felt coverings and deal with cracking as soon as possible; a leaking roof will allow water to seep into the structure of the house, accelerating the rotting process and generally dampening the environment inside.

Patch repairs aren't generally recommended and it's usually best to replace the complete sheet of felt. It will need to be fitted correctly and carefully to ensure a weather-tight seal. An easier but more expensive option can be to buy a replacement roof panel from the house manufacturer, if such spares are offered.

It is vital to ensure that the roof of your hen house is kept intact at all times. A leaking roof will lead to a damp and rotting house – not a pleasant environment for your hens.

HANDY REFUGE

Although roofing felt has been successfully used for covering roofs of garden sheds since Adam was a boy, its use on chicken houses is potentially more problematic. Not only does it crack and leak with age, but it can also offer a wonderfully secure shelter for one of the most dreaded poultry pests: red mite.

18 The hen house walls are damp at night

CAUSE

Inadequate ventilation can lead to a damp atmosphere in the hen house at night, as the heat and moisture generated by the sleeping birds turns to condensation. This is a serious problem, and can be especially damaging if the unit is over-stocked.

SOLUTION

The fact that chickens have a naturally high body temperature (typically 40–41°C), but don't have the ability to sweat as a means of cooling like we do, makes the quality of the environment in which they're housed – especially overnight – all the more important. They use panting and feather adjustment to lower their body temperature, so the air they have to breathe inside their house needs to be both fresh and dry. Insufficient ventilation, coupled with a lack of overall air volume inside the house, are two of the main reasons why hens can become stressed and even ill.

The trend nowadays for small, manageable hen houses is great from a convenience point of view, but doesn't always work for the birds inside. Designs featuring plenty of headroom for the birds are generally best, as are those with a good number of ventilation holes or slots at the top of the structure. This will ensure that the heat and moisture generated by the birds as they roost at night will be drawn up, away and out.

Some keepers cover or close ventilation points during the winter, imagining that they are doing their birds a favour by helping to keep them warm. In fact, they're usually doing nothing but harm – the hen house needs to be well ventilated, even in winter.

ADEQUATE VENTILATION?

Hen house manufacturers who are inexperienced or who want to cut costs will often ignore or overlook the basic needs of hens, resulting in sub-standard structures that are either under-ventilated or let draughts into the house. When corners are cut house design suffers and so, ultimately, do the hens forced to live in these houses.

Converted garden sheds can make excellent hen houses; the volume of air inside the unit works brilliantly to help guard against damaging dampness and humidity, assuming proper ventilation is provided with drillings at the top of the walls, just below the roofline.

19

The run floor in my hen house/run combination unit has gone muddy

CAUSE

If you don't relocate hen houses that have attached runs on a daily basis, then the ground will be ruined.

SOLUTION

Hen house/run combination units are great for convenience, and for keeping hens contained (especially in urban environments). However, their success relies on correct stocking density, regular movement onto fresh ground and/or operating a 'deep litter' system on the run floor.

When keeping hens in this sort of restricted space it is especially important to avoid overcrowding. It is vital that the birds are comfortable, content and unstressed in their contained environment, so always stick to the house manufacturer's guide about bird capacity or, better still, set the number just below the suggested limit.

If you're standing the unit on grass, make sure that it doesn't remain in one place for more than a day or two. In a matter of days even a small group of hens will damage the root structure of lawn grass beyond recovery by pecking and scratching at it. If you're unable to move the unit on to fresh ground this frequently, then you'll need to layer the run area with a thick covering of bedding material (ideally, wood shavings on a paving slab or concrete base), and protect it from the rain with the addition of appropriately positioned side panels. This layer will need to be deep enough to allow the birds to scratch, and for their droppings to be effectively absorbed if they're not being collected. This sort of 'deep litter' approach will require careful management to ensure good health and welfare, and it will typically have to be completely replaced once a year.

ARE THEY AMUSED?

Chickens confined to a fixed run should never be allowed to become bored. If you can't move the run regularly on to fresh ground, then ensure the birds have logs or a hay bale to perch on, suspended greens to peck at and a regular scratch feed.

Allowing your hen run to become muddy will be inconvenient for you and unpleasant for your birds.

My hen house blew over

CAUSE

The most common reasons why a hen house might blow over – discounting freak weather conditions – are that it has been poorly positioned or badly designed and made.

SOLUTION

The placement of modern lightweight hen houses is an important consideration. If your property is in an exposed location, you will need to shelter the house from the prevailing wind or, if this isn't possible, anchor it using stakes and brackets.

Think about the roof design, too. Units built to a pent-roof style (with a single, flat-panel roof set at an angle) are best orientated so that the prevailing wind hits the low end of the roof first, so that it presses the house down into the ground as it blows up and over. Pitched roof designs are less sensitive in this respect, although it is still advisable to angle them so that the prevailing wind is hitting one sloped roof panel, rather than the flat-sided front or back panel.

Hen houses need to be raised appreciably above ground level (20cm) to discourage rats and mice from nesting underneath, which can reduce stability further by raising the centre of gravity. Again, the best advice is to site your hen house where it will be sheltered from wind. Also, buy the best that you can afford, made from decent timber that is at least 10mm thick. If you are still concerned, drive in wooden fence posts or metal stakes nearby, and use them to anchor the structure as best you can. Paving slabs or house bricks can be a useful temporary measure to weigh down hen houses during stormy weather.

BEWARE A BARGAIN

The popularity of chicken keeping as a hobby has prompted something of a revolution in hen house design. It is getting hard to keep up with all the hen house producers on the market, and there is a vast difference in the quality of their outputs. With increasing numbers of producers determined to drive down selling prices, short-cuts inevitably get taken; some are even being sold now as self-assembly, flat-packed units.

Always take care with the siting of your hen house, to ensure that it is protected from the worst of the weather. Locating a hen house close to fencing or other natural wind-breaks, and among trees, will usually prove advantageous.

CHAPTER THREE
CHICKEN RUNS

For many chicken keepers, the run can be something of a compromise. Space, especially for those keeping hens in an urban environment, is one of the most obvious constraints. Be realistic about the suitability of the area you have for keeping chickens. Runs need to be appropriately sized for the type and number of birds, as well as large enough to withstand the wear and tear caused by the birds' daily activity.

Chickens will happily utilise as much space as you're prepared to give them, and in most residential back gardens, this tends to be at a premium. Other garden users need to be considered, too: not everyone in the family will want the entire garden turned over to the hens, or enjoy the sight of a hen house while relaxing on the patio. The popular dark-coloured electrified poultry netting is a reasonably unobtrusive, very effective and easily movable solution for keeping a few back-garden hens happy and safe; the sort of permanent fencing required for a larger, fixed chicken enclosure will be a good deal more utilitarian to look at.

Security is a key concern when building a chicken run; if you opt for electric fencing, you need to be aware of the problems and pitfalls involved. Various steps can be taken to ensure that your hens feel safe, sheltered and have plenty to keep them active and entertained.

21 My birds keep escaping

CAUSE

The most common reasons that birds get out of their run (assuming the gate hasn't been left open!) are that there is inadequate fencing to contain them, or that they simply fly up, up and away.

SOLUTION

Most chickens aren't great fliers. As a general rule, bantams are usually better at it than their large fowl counterparts, while the 'light' breeds (typically those with Mediterranean roots, such as the Leghorn, Spanish and Minorca) can be expected to fly better than the larger, 'heavy' breeds, and are generally more excitable, too.

Nevertheless, even the worst fliers are normally perfectly capable of short bursts of fluttering, which can be sufficient to get them up and over the standard, 1m-high electrified netting used to enclose many domestic chicken runs. To prevent this, the fences obviously need to be made higher, which means they'll also need to be permanent.

This, of course, ushers in a whole new level of commitment, in terms of the set-up required, the cost involved and its visual impact. To be effective at containment, a permanent fence will need to be at least 2m high and constructed of sturdy wire mesh supported on properly dug-in posts. The corner posts must be the strongest (ideally braced) and mounted in concrete; they will provide the strength and tension in the fence as a whole, and so must be of a larger diameter. Electrification will also need to be considered as a rigid wire fence will be easily scaleable by a determined fox (they are superb climbers). Typically, single-strand, powered wires around the base, halfway up and at the top, are the preferred anti-fox measure with permanent fences.

Chickens naturally feel nervous and exposed on open ground, so will seek the highest perch they can at the first sign of trouble or at dusk.

22 A fox took my chickens

CAUSE

Fox attacks most commonly result from inadequate fencing, failed electric fencing or the electric fence being left switched off altogether.

SOLUTION

Electric fencing shouldn't be regarded as the definitive answer to the predator problem. These systems require careful and methodical maintenance and attention if they are to remain effective. Foxes have territories that they patrol on a regular basis. If this happens to include your garden and chicken run, they will check the status of the electric fence each time they pass. Foxes seem to have an innate ability to detect whether or not the power is on, and without any current pulsing around the system, a fox will be able to jump straight through and into the pen.

Power failures are always going to be a risk with battery-powered systems. Ideally, you should keep a second, charged battery in reserve so that failures can be put right as soon as possible. The battery obviously has a finite life: make sure you find out what it is so that you can plan ahead for changeovers, thus minimising the risk of unnecessary and potentially dangerous downtime. If the fencing system short-circuits (due to unforeseen contact that allows current to drain away to earth), this will dramatically shorten how long the battery holds its charge, so this needs to be avoided.

Regular checks on the voltage levels running through the fence, using one of the many types of tester now available, should become part of your good husbandry routine. To remain an effective predator deterrent, an electric fence needs to be pulsing at the required 5,000–6,000V. Voltages appreciably lower than this – caused by a faltering battery – will be detectable by prowling foxes, and may tempt them into jumping over.

23 The electric fence battery failed too quickly

CAUSE

There are a number of reasons why this may happen, including use of the wrong type of battery, inadequate charge in the first place, an ageing battery or electrical leakage caused by accidental short-circuiting.

SOLUTION

Battery management should be high on your list of priorities. Buy yourself a voltmeter so that you can keep a regular eye on the state of charge within the battery, and also take care to charge it properly each time. Many modern chargers offer a 'fast charge' option, but this is best avoided in anything other than an emergency situation. It is always better for the long-term life of the battery to charge it slowly, with the charging unit set to normal.

Don't run your electric fence using the old car battery that you found in the garage. This will work well enough initially, but batteries made for use in automotive applications aren't designed for the sort of constant, deep-draining use required of those hooked up to a fence energiser. When purchasing your battery you need to specify a 12V leisure battery: the sort intended for use in caravans and golf buggies.

Battery age can play a part, too. Just as the unit on your car will typically fail after four or five years, those used for electric fencing won't last forever either. A typical sign of failing performance is a shorter working life between charges, so be aware of this and start planning to buy a replacement.

The chances of electrical leakage increase the longer the fence is, simply because of the amazing speed at which weeds and other vegetation can grow during the spring and summer, and the greater distance that needs to be checked. It is important to walk the entire fence line every other day to keep an eye out for potential problems.

24 The fence voltage is too low

CAUSE

Reductions in the voltage circulating around an electric fence can be triggered by a faulty or inadequate energiser, poor earthing, dirty connections or a failing battery.

SOLUTION

An electrified fence is only as good as the electricity pulsing down its wires. If this falls below par, then so will the effectiveness of the fence. One of the most important things with any electric fence system is to make sure that the energiser unit is able to cope with the length of fence. People run into problems when they extend their run size by adding an extra length of netting or two, then expect their existing energiser to up its game accordingly. These units have a specific output limit.

It is also worth bearing in mind that the quoted power output – and the recommended length of fence it is intended for – is for operation under ideal conditions. However, we all know that out in the real world, 'ideal conditions' very rarely occur, so pushing performance to the limits of the stated output is asking for trouble. With energisers, it is always better to have a little in reserve to help battle against the inevitable electrical inefficiencies that are bound to creep into the system with time.

Longer runs of fencing may also benefit from additional earth stakes to improve operational efficiency. As with any electrical circuit, a good connection to earth is essential otherwise no current will flow. With electrified netting, this earth is typically provided by a 1-m long metal spike, which needs to be pushed all the way down into the ground to ensure the best operation. In addition, all connections between netting lengths should be regularly checked for tightness and cleanliness of contact.

SHOCK TACTICS

Don't be put off by the idea of thousands of volts buzzing around electrified poultry netting; it's not as scary as it sounds. These systems are widely used in domestic situations. Quite simply, electrified netting provides the most flexible and effective anti-fox measure there is.

A fence tester is an essential piece of kit for monitoring the level of voltage that is pulsing through your fence. The wires need to pack a decent punch to provide a worthwhile deterrent.

25 I keep hearing the electric fence clicking

CAUSE

This sound indicates that your electrified fence is short-circuiting and, therefore, running inefficiently.

SOLUTION

Anything that comes into contact with your electric fence, such as nearby vegetation, will provide an easy route for the electricity to take to earth, causing the system to short-circuit. Grass and weeds that grow to over 20cm close to the base of the fencing will present a short-circuit risk, especially during spells of wet weather, or even when there's a heavy dew. Under these conditions it is perfectly possible to hear the regular clicking as the current jumps between the fence and the wet greenery or branch that's touching it. Any significant dip in the voltage running through the wires of your fence could provide an opportunity for a passing predator to gain entry. What's more, if it's a battery-powered system, the working life of the battery will be greatly reduced by short-circuiting.

So, it is essential that all growth is kept well controlled at the base of all electric fencing. The most effective solution is to run a strip of weed-suppressant material around the base of the fence, although this will have to be moved if the enclosure is relocated. Weed killer is another option, but great care must be taken with the use of toxic products near your chickens. Most keepers tend to prefer keeping weed and grass growth in check with the regular use of a rotary grass cutter or strimmer. This will involve moving the fence to one side as you work; enlist a friend to help with the fence as you cut.

The wild grass on the outside of this poultry run has been allowed to get out of hand, rendering the electrified netting inoperable. The effectiveness of these fences relies on all foliage, stalks and low-hanging branches being kept well away from the electrified wires.

26 There are holes appearing in the chicken run

CAUSE

Holes inside the chicken run are most likely to be caused by the birds themselves, but they could also be made by rats (if made by the latter, they will be much smaller). Holes found around the outside of the run will usually have been made by a fox or a badger, trying to dig under the fence to get at the birds inside.

SOLUTION

Hens are surprisingly good at digging, and typically they do so in an attempt to create dust baths for themselves. Consequently, you'll often find holes in the driest parts of their enclosure, under the hen house or at the bases of hedges or bushes. As long as the number of holes doesn't get out of hand and start impinging on the grass cover inside the run, this doesn't pose much of a problem. If it worries you, try adding a man-made dust bath for them to use.

The presence of rats is definitely a cause for concern. Colonisation beneath an old-style hen house that sits directly on the ground can be highlighted by the appearance of small, round holes in the ground close to the base of the house. Not only can rats spread disease, but they'll also cause structural damage to the hen house – often creating water leaks and draughts – and can take eggs and even young chicks, given the opportunity. It's vital to be responsible about your anti-rodent measures.

Scrapings and other signs of ground disturbance and digging on the outside of your chicken enclosure fence are a sign that predators such as foxes, badgers or mink have attempted to access the run. If they have, it is likely that your electrified fencing isn't providing enough of a deterrent. The voltage may have dropped due to problems with the battery or short-circuiting somewhere in the system.

Holes inside the chicken run will most likely have been dug by the hens, as they search for tasty roots, insects or simply create a dust bath. However, scrapings on the outside, at the base of the fence, are a cause for concern.

27 My hens are huddling next to the hedge

CAUSE

Chickens need overhead cover, both for a sense of security and for shelter from strong, summer sunshine. If their run is bare, then they'll do the best they can to find shelter, which will mean gathering at the base of a hedge if that is all that's available.

SOLUTION

All domestic chickens feel happiest and most relaxed when they have decent overhead cover. They naturally fear predators from the sky, such as eagles or other birds of prey, so it is very important for their peace of mind that you make some form of shelter available in their run. Ideally this will be natural, in the form of large bushes or trees.

If this isn't possible, then simple field shelters are easy to make using straw bales, sheep hurdles, sheets of Onduline or plywood. You don't need to design and build a complex structure, just something airy but sheltered, and large enough for all the birds to use at once if they wish to. You can add some rudimentary perching, and they'll also appreciate enough space to create a dust bath.

The sort of dry soil typically found at the base of many garden hedges can make these areas favourite dust bath sites for hens, and they'll happily scratch away, creating holes in which to carry out this natural behaviour. If you don't want this to happen, then you'll need to give them an alternative. Provide a large, low-sided box that's half-filled with a mixture of dry soil, clean builder's sand, wood ash from a traditional bonfire and food-grade diatomaceous earth. Dust bathing is an important part of daily life for a free-ranging chicken, and it helps them to counter parasite activity from the likes of lice, mites and fleas. Offering your birds the opportunity to take a dust bath is an important aspect of helping to ensure that their quality of life is as good as it can be.

FREE-RANGE FOLLY

One of the most disconcerting things that you can do to a
chicken is to plonk it down in the middle of a large, open field.
Being descended from jungle fowl, the lack of cover and shelter
will prove extremely stressful for chickens. What's more,
such action can prove detrimental to their overall health
and laying performance.

*Chickens will huddle
under anything that
provides a source of shelter
in the chicken run.*

28 My hens seem bored

CAUSE

Small, featureless runs, with minimal vegetation and no other distractions, can cause lethargy, depression and even ill health among chickens.

SOLUTION

Chickens are generally sociable, inquisitive and active creatures. Their natural instincts are to roam and to forage for food as they do so; confining them does much to curb their natural inclinations. For this reason, it is very important that keepers do all they can to make what space the birds have as interesting and stimulating as possible. The smaller the area, the more essential such environmental enhancements are.

You needn't spend a fortune to make a great difference to the everyday lives of your birds: a bale or two of straw, a few large logs or some additional perching are all simple additions you can make. Fresh raw greens like cabbages, lettuces or sprout stalks hung overhead will provide great and nutritious pecking targets. If you suspend these at just above head height, so that the birds have to stretch and then jump a little to get at them, this will help to keep them exercised and in condition, too – another important consideration for birds that are cooped up in a relatively small space. Likewise, you can hang old CDs on strings in the pen – the way they sparkle in the sunshine will certainly attract an inquisitive peck or two.

When clearing flowerbeds or vegetable patches, don't forget that your chickens will love to peck at the freshly dug stalks and roots, especially if there is the odd worm or two mixed in. Even fresh pieces of turf provide a great distraction, as will an opened growbag full of tomato plant roots.

Keeping your hens active and amused can be simple and cheap to do; suspended fresh greens create a great and healthy diversion.

29 My chickens are eating my vegetables

CAUSE

Chickens, if not prevented from doing so, will happily scratch and peck their way through herbaceous borders and vegetable patches, and they'll do it surprisingly quickly, too!

SOLUTION

Most breeds of chicken can be very destructive if allowed to roam freely through a garden, and they will have no respect for specimen shrubs, rare plants or a neatly tended cabbage patch. They will simply have a go at whatever they fancy, given the opportunity. If you want to protect special plants and vegetables from free-ranging chickens, your only options are to control the activity of the birds, or to protect individual plants and borders. Creating a 'no-bird' zone is the most effective, assuming you can keep the feathered friends out. Don't forget that a determined chicken with rich pickings on its mind can flutter high enough to get over most low fences and garden hedges, so any partitioning will need to be fairly serious (at least 2m high). Clipping the birds' wings may be the easiest solution, as long as you're not intending to exhibit them.

Alternatively, you can make or buy fruit cages to fit over fruit bushes and vegetable plots; this will be relatively easy to do if you grow in raised beds. However, don't forget the benefits of allowing your hens to pick over your vegetable beds at the end of the season. They'll relish the chance to get at the resident insect life, and will effectively clear weeds and fertilise as they go. Some keepers even contain their birds for a day or two on each bed, to ensure a thorough going-over.

SCRATCHING BEHAVIOUR

If you are worried about the hen damage to precious flower borders, then take a bit of care with breed choice to limit the potential for damage. As a general rule, the feather-legged breeds tends to scratch less than those with clean legs; bantams, too, tend to be less destructive than large fowl.

While hens can be great garden friends and perform very useful insect-clearing duties, they will also run riot through precious herbaceous borders and vegetable beds.

30 I'm unsure whether my run needs a roof

CAUSE

Too much rain, poor surface quality, excessive interest from wild birds or the need for some shade can all be reasons why you may wish to cover all or part of your chicken run with a roof section.

SOLUTION

If you are struggling to maintain the quality of your run floor and you're unable to switch to a different location or extend the run to give the birds more space, then one relatively simple solution can be to add a roof. Not only will this provide the run floor with some protection from the elements, it will also provide a useful amount of shade for the occupants during sunny conditions and protect them from the worst of the winter weather, while still allowing them to get out of their house. If all you want is to keep wild birds out (and your own birds in), simply stretch a suitably sized section of netting across the area.

If you decide that a more substantial roof is needed, remember that it will require decent support, certainly around the edges, and probably across the middle, too. If you already have a run made from properly installed, permanent fencing, then the existing posts may be strong enough to double up as supports for a roof section. If not, then you will have to start from scratch, with corner posts and a roof framework. Onduline, which is a bituminous, corrugated sheeting material, can represent a durable roofing material, although it is quite expensive. Corrugated iron is a cheaper but less attractive option; it's certainly functional, quite light and not bad at supporting its own weight. You could also opt for clear, corrugated plastic, or a combination of this and a 'solid' version. It all comes down to budget, and the time you want to devote to the project.

THE WIRE

Wire quality matters with regard to permanent chicken runs. Don't imagine that you are going to enjoy much peace of mind using appropriate-sounding chicken wire. It is simply too weak; a determined rat, fox or badger will bite through it. Securely mounted, sturdily fixed, rigid weldmesh, although unsightly, is the only answer.

Putting a roof on your chicken enclosure will ensure peace of mind if you are keeping flighty birds in an urban environment and effective containment is a requirement.

CHAPTER FOUR

RODENTS AND OTHER PESTS

People who do not keep chickens are often quick to criticise those who do because of the perception that the presence of hens in the back garden will attract rodents and, in particular, rats. The reality is that rats are everywhere already. Anyone who has a compost heap, log pile, garden shed or building materials stored outdoors is likely to be providing sanctuary to these versatile and enduring creatures.

Of course, there are plenty of mistakes inexperienced chicken keepers can make that will increase the likelihood of rats taking an interest in their chicken set-up. But, with a bit of forethought, some good planning and by adopting a decent husbandry routine, there is absolutely no reason why chickens kept in the back garden should provide a specific attraction to rodents. Well-kept hens shouldn't attract mice or flies either and, in fact, their efforts at tackling snail and slug populations should be applauded by every gardener in the locality.

It pays to be prepared for any eventuality, though, and this chapter covers some of the challenges involved in getting rid of pests like rats should they become a problem, as well as practical tips on how to discourage predators such as foxes and birds of prey from paying your chickens an unwelcome visit.

31 My chicks are disappearing

CAUSE

It is always tempting to let young chicks out into the chicken run on warm, sunny days. However, these youngsters (and, indeed, older chickens, too) can be vulnerable to attack from birds of prey.

SOLUTION

Flying predators can be a good deal more difficult to deter than the four-legged variety, especially if you allow your hens free range over a large area. Those in quiet, rural situations will always face a greater potential threat from birds of prey, but do not discount the risk just because you live in a town or city. Various types of hawk are increasingly sighted in urban locations; the plentiful supply of pigeons has much to do with this.

The only sure-fire way to stop attack from the air is to install top cover over your chicken run, but the practicality of this is very much dependent on the run's size. Roofing large areas is just not viable in most cases, but you can ease your hens' level of exposure by adding field shelters: either natural (shrubs and bushes) or man-made. Bird scarers, of the shiny, flappy type, can be useful too, as can a good old-fashioned scarecrow. Also, the more time you can spend in the run with your birds, the better. Predators will be wary of any human presence. Unfortunately, if a couple of hawks happen to start nesting nearby and get to know about your chickens, then regular strikes are likely. You could introduce a cockerel to your flock, as male birds are great at acting as lookouts. They will issue a warning call to their females if they spot an overhead threat, alerting the flock to take cover.

At dusk and after dark, the threat switches to one of owl attack. Once again, much depends on your location and the prevalence and species of owls there. But if you know they are around, be sure to get your hens securely shut in their house as dusk falls.

A scarecrow could help to ward off opportunistic birds of prey.

32 The bait station is not working

CAUSE

It is no good putting down rat poison if none of the rats eat it. Rats are naturally very suspicious of anything new, so they are likely to avoid a shiny, new bait station, even if it is loaded with the most tempting treat.

SOLUTION

The local rat population will need to get thoroughly used to the bait station before they will be relaxed enough to venture inside. Moving it around will not help, because each time you switch locations, you will simply reset the 'acceptance clock' to zero. So, pick your site carefully, place the bait station and then leave it. Rats typically avoid crossing open spaces, preferring to stick close to the edges of buildings, garden sheds, fences, and so on. There is therefore never any point in placing a bait station out in the middle of a chicken run or other large area; it has to be located right up against a boundary of the sort already described.

You need to think like a rat and take some time to look for signs of 'rat runs'. If you have a significant infestation, then it is quite possible that you will be able to spot the regular routes taken by rats during their night-time excursions; look for droppings and trampled vegetation running close to the walls and boundaries. The idea is then to place the bait station directly on one of these well-used rat runs as, that way, you can be pretty well guaranteed that the rats will encounter it on their travels.

Another trick worth employing is to 'age' the bait box before setting it. Experienced keepers recommend smearing the new bait box with mud and leaving it outside for a couple of weeks before setting it, so that it becomes thoroughly weathered and loses its 'new' smell and look. This is the only way to ensure that rats will regard it without suspicion and, therefore, begin going inside to take the bait.

Bait stations intended to control the local rat population need to be well designed to protect other pets and wildlife, and carefully positioned.

SENSIBLE MEASURES

Always wear gloves when handling, checking or replenishing an operational bait station, not only for your health and safety, but also to minimise the chances of you transferring any of your own smells to the unit. 'Human' scents such as soap, deodorant or aftershave, as well as sweat and the natural oils from your skin, can all be off-putting for rats.

33 I want to get rid of rats in a humane way

CAUSE

Some people dislike the use of rodent poisons, because they object to the typically slow way in which they act to kill the rats that consume them. Fortunately, there are other ways to control rampant rodents, but welfare can remain an issue.

SOLUTION

Traditionally, farmers used terriers to flush out and quickly kill rats but, nowadays, this is not a practical solution in most domestic, back garden situations. An alternative approach for many hen keepers is to use spring-based traps, with some form of hinged arm that is designed to snap shut and break the rat's neck when triggered by weight on a pressure plate.

There are many variations on this theme available nowadays at a range of prices. However, it should also be pointed out that the use of these devices is not always as clear-cut as you might imagine. Everything depends on the falling arm striking the rat or mouse on the neck, behind the head; this is the only way that a 'clean kill' can be guaranteed. The chances of this happening rely, almost entirely, on the orientation of the animal as it attempts to take the tempting bait; it needs to be pointing headlong into the trap for the most efficient operation.

Unfortunately, this cannot be guaranteed. Animals that attempt to take the bait from the side or that, perhaps, get pushed onto the pressure plate without realising it, can get caught by the leg or some other part of the body. This can lead to a slow and agonising end that nobody with an ounce of sensitivity would like to see any animal suffer. But it is always a risk when using these traps, and discovering a badly injured and trapped rat is not only extremely unpleasant, but also demands further action to put the animal out of its misery. So, think carefully before embarking down the rat trap route.

REGULAR CHECKS

If you are using snap traps against rats and mice, it's vital to check them at least once a day, preferably more often. Even when aiming to catch and kill vermin, you have a specific responsibility to minimise any suffering caused. Animals caught but not killed outright must be despatched without delay.

Snap traps like these do not always result in a clean kill. If this worries you, think carefully before using them.

34 I am worried about contracting Weil's disease

CAUSE

Weil's disease is a very unpleasant condition in humans that can be caused by direct contact with rat urine. Mercifully, cases remain relatively rare in regions with temperate climates; it is more common in tropical and sub-tropical areas.

SOLUTION

Also known as leptospirosis, Weil's disease is a type of bacterial infection that is most commonly carried by rats (and cattle), although the presence of the disease in rats (held in their kidneys) does not cause any ill effects in the rodents themselves. While bites from infected rats are unlikely to transmit the disease, humans can contract it by contact with an infected animal's urine. Typically, this happens when contaminated water enters the body via a break in the skin. Weil's-carrying rats urinating in your hens' drinkers can pose a very real infection risk.

Symptoms are flu-like in the early stages. In mild cases, sufferers will recover after five days, although periods of fatigue and depression may follow. More serious cases will trigger a second phase, when the flu-like symptoms return, but more severely. Jaundice may also result, together with diarrhoea and, in really serious cases, liver, kidney, respiratory system and even heart failure may occur. So, as you can appreciate, Weil's disease is not something to be taken lightly.

The risk, although small in many cases, remains a very real one, and anyone dealing with rat infestations under their hen house, or clearing piles of rubbish from the garden that have been left untouched for long periods of time, should take extra care. Always wear gloves when dealing with potentially contaminated water containers, ensure any cuts and scratches are covered with waterproof plaster, and wash your hands thoroughly with soap and hot water afterwards.

Always wear gloves when there is a risk of coming into contact with anything that may be contaminated with rat urine, such as chicken drinkers, especially if you have cuts or scratches on your hands.

35 I have found a hole in the hen house floor

CAUSE

Much depends on the appearance of the hole. Does it look as if it has been caused by wood rot? If so, a water leak is most likely at the root of the problem. Does it look as if the wood has been chewed or gnawed at? If so, it could very well be the work of determined rodents.

SOLUTION

Dealing with a leaking hen house is generally a relatively straightforward matter, once you have isolated the source of the problem. Damage to the roofing felt, warped panels, construction issues or – rather more fundamentally – shortfalls in the design of the unit can all lead to water finding its way in. If the leak has been bad enough to result in a rotten house floor, then it certainly will not be a recent occurrence; it will have taken many months for this type of problem to develop. In the case of such a long-standing issue, staining will make the route the water has been taking to reach the rotten point of the house floor clearly identifiable and it will be easy enough to trace the path upwards and find the entry point, so that appropriate repairs can be made to the roof and floor.

Rodent damage is usually more straightforward to repair, given that you are only usually talking about a single hole; a basic patch that is screwed or glued into place should be all that is necessary. Colonies of rats are probably best and most effectively dealt with by your local pest control expert. After the infestation has been cleared, to avoid the problem recurring it would be prudent to raise the hen house on legs or blocks (to at least 20cm above the ground). This will ensure that plenty of natural light reaches the ground underneath – enough to make it a thoroughly unattractive proposition to any rats searching for a new home. If you are reluctant to do this, then put down a pad of concrete or heavy, close-fitting paving slabs on which the house can stand thereafter.

Holes like this one, where the wood has been gnawed and chewed at, are most likely caused by rodents such as rats.

RAT PARADISE

If rats are the culprits, in most cases they will have colonised the ground immediately beneath the house: something that happens with houses that sit directly on the ground, or are raised just a few centimetres above it. The shelter, seclusion and warmth provided by the house in this sort of situation offer just about everything a rat could want and, with feeders typically close by, there is a very convenient source of food on tap, too.

36 I have caught a live rat; what now?

CAUSE

The use of so-called 'live traps' is regarded by many people as the most humane form of rodent control of all and, on the face of it, they do seem less unpleasant than poison or spring traps. However, they can throw up many more problems than they solve.

SOLUTION

There are two primary issues associated with the use of live traps, and both relate to animal welfare. Although attitudes and regulations vary around the world, the belief in much of Europe and America is that trapped animals must be treated humanely and not exposed to undue suffering. Holding a wild rat in a confined, cage-like live trap for any length of time will prove extremely stressful for the animal, so this type of trap must be checked on a frequent basis.

But the big problem, of course, is what to do with the rat once you have caught it. If you are unable to shoot it there and then, what are the alternatives? Other options considered to be humane methods of disposal are a sharp and accurate blow to the head with a suitable instrument (very difficult, because the angry, scared rat will need to be removed from the trap first) or a lethal dose of gaseous or injectable anaesthetic (impractical in the domestic environment). Drowning is considered inhumane, as it causes unacceptable suffering, and releasing the rat is not recommended because it simply transfers the problem to somewhere else. Doing this has the potential to create health concerns for people, pets and other domestic animals at the new location, as well as upsetting native ecosystems.

So, unless you are willing and able to shoot the rat you have caught, then the use of live traps really is best avoided.

Catching a rat in a humane trap is only half the solution; there remains the question of what to do with the captured rodent.

37 Something has been digging under the fence

CAUSE

If you find evidence of digging at the base of your chicken run fence, it can only mean one thing: a predator is trying to get at your chickens. By enclosing chickens in a confined space in what typically amounts to very unnatural conditions, we are effectively putting them on a plate for hungry, determined predators – unless adequate defence measures are put in place.

SOLUTION

For many chicken enthusiasts, the fox represents public enemy number one; it is an animal that people just love to hate. The fox is a natural predator, a superlative opportunist and a great survivor. For as long as we fail to protect the vulnerable, flightless chickens we corral in our back gardens, foxes will go on taking what they can, every time they get the chance. It is what they do, and is the harsh reality of nature. Those at fault are the keepers who fail their birds by providing inadequate anti-fox measures. Foxes (and badgers) are excellent and tenacious diggers and, if they think it is worthwhile, they will scratch their way under a chicken run fence in next to no time.

The only real way to prevent this from happening is to create a physical barrier to block their path, by extending the base of your wire fence. You can do this straight down into the ground; you will need to go in at least 60–80cm to be really effective. Alternatively, you can extend the wire mesh out horizontally, away from the base of the fence and on top of the ground. This is obviously the easier option as it requires no digging, but is nevertheless effective. The predator's natural inclination is to start digging at the base of the obstacle; securely pegging down a 100cm-wide apron of wire mesh should prevent this effectively.

If something has been digging at the base of your chicken run fence, then it is time to increase your security.

IF AT FIRST...

Predators like foxes and badgers are nothing if not determined. Never underestimate their tenaciousness, or doubt their ongoing efforts to get at your chickens. They will be perfectly happy to play the long game, waiting for you to relax your guard or make a mistake, before they pounce with typically devastating results.

38 Why has a fox killed so many of my hens?

CAUSE

Foxes are great opportunists and will take food whenever they can. Coming across a pen full of plump chickens is a chance too good to miss, as far as they are concerned, so they will kill all they can, when they can.

SOLUTION

The only solution to this problem is to keep the fox out of the chicken run by ensuring that your defences – be they electrified or permanent, rigid wire fencing – are up to the job. Electric poultry netting is only as good as the current that is pulsing through it; if the supply weakens due to vegetation-triggered short circuits or a dying battery, then a hungry fox is not going to think twice about hopping over to get at your hens. Persistent, battery-related problems could indicate that the time is right to consider a switch to a more reliable mains-powered supply.

The reason that foxes tend to kill so many chickens in one attack is that their intention is to take the carcasses away afterwards to a food store somewhere else. Many people like to think that foxes simply enjoy killing, which is why they so often appear to kill so many more birds than they could reasonably consume: another factor that fuels the widely held anti-fox sentiments. The reality, though, is that the fox rarely gets enough time to take all the dead birds away to its food store. Obviously, the birds can only be moved singly and, depending on where the food store is, this can take some time. The devastation in the chicken pen is often discovered by the birds' keeper quite soon after the initial attack and, because the fox may have only had the chance to take one or two birds away, the scene of carnage remains a typically shocking one.

The fox is a master predator and will take advantage of any opportunity it finds. Chickens corralled in an inadequate, back garden pen can be an easy target.

39 I am scared of using poison

CAUSE

There are many types of rat poison available in the shops but, understandably, a lot of people are wary of using such products.

SOLUTION

The use of poison is never a pleasant business but, for the chicken keeper, it is one of the necessary requirements of running a well-organised and responsibly managed set-up. However, the products involved need to be selected carefully and used prudently. Many chicken enthusiasts keep other pets, too, so the last thing they want is to endanger those animals, or to put the local wildlife at needless risk.

The key is to follow the manufacturer's instructions to the letter. These will vary depending on the product, but doing things as instructed, and while wearing the appropriate protective clothing, will ensure that safety is maintained at all times. The basic precautions are essentially common sense: avoid the product coming into direct contact with your skin and eyes, and wash your hands carefully once you have finished.

In terms of protecting pets (including your chickens) and other wildlife in your garden, the vital requirement is that access to the poison is effectively limited so that nothing else can reach it apart from the intended target. If you're using poison that is designed to be dropped down a rat hole, then make sure that after application you cover the hole with a heavy brick or paving slab so that nothing else can access the poison from above. Bait stations are carefully designed so that the poison is well-insulated from the natural environment. Entry points are sized according to the pests being targeted – in this case, rats and mice – which means that the poison you place in it will not be available to other creatures such as hedgehogs, wild birds, chickens, cats or dogs.

40 The hen house has been invaded by rooks

CAUSE

Rooks are very clever birds and great opportunists when it comes to feeding. They will take eggs from inside the hen house once they learn they're in there.

SOLUTION

It doesn't really matter what sort of hen house you have: if the local rooks realise there are rich pickings inside in the form of delicious, fresh eggs, they will quickly work out how to get in and take them.

What's more, it won't be a one-off event, once the thieves realise that there is a regular supply of eggs to be enjoyed. Fortunately, this sort of destructive behaviour tends to be restricted to spring and early summer. The rooks will usually work in pairs, and preventing them from entering the house is the name of the game.

The best option, of course, it to keep the pop hole closed until you've had a chance to collect the eggs, but the practicality of doing this rather depends on how early the hens lay. Keeping them cooped up for too long – especially when the weather is warm – is never a good idea.

Other deterrents, such as partially covering the open pop hole with a hessian curtain, or hanging up old CDs near the entrance to the hen house to turn and sparkle in the breeze, can have some effect. It might also be worth moving the hen house to a new location, if you've got the room to do so. Rooks, though, are canny operators, and are always happy to play a waiting game. They will always be scared away by a human presence, but will typically circle around, then settle in a nearby tree at a safe distance to watch until the coast is clear.

CHAPTER FIVE

PARASITES

All hen keepers will have to deal with parasites sooner or later. The trouble is that these little pests are not always easy to recognise, so many novice keepers fail to appreciate there is a problem until things have become quite serious. Parasites can be either internal or external (living inside or on the bird), although red mite is an exception because it feeds on the chicken at night, but lives within the structure of the hen house during daylight hours.

Getting to grips with parasites at an early stage is the key to success, and doing so involves developing an understanding of what you are up against and how best to tackle it. It is also important to appreciate that with adversaries such as the red mite, winning the battle outright is almost impossible. Once these tiny creatures have become established in a hen house, the best that most keepers can hope for is to keep their numbers at a manageable level.

Dealing with parasites demands a methodical, organised approach. Rarely will one treatment do the trick and, in most cases, repeated applications of your chosen medication will be required. Many of these treatments are straightforward and can therefore be performed by the chicken keeper. Chickens are hardy creatures, and with the right care and attention they should soon be on the mend.

41

The hens are reluctant to enter the hen house

CAUSE

There are a number of potential reasons for this, including bullying and peer pressure within the group's pecking order, but one of the most potentially serious is a red mite infestation.

SOLUTION

Hens can cope with low levels of red mite activity but if parasite numbers reach infestation levels (which can easily and quickly happen unless the pests are tackled effectively), then every night can become a tortuous ordeal for your hens. So much so, in fact, that the birds will become reluctant to enter the hen house at night because they know full well what is waiting for them. But the consequences for the chickens are potentially more than just irritating and uncomfortable. Large numbers of red mite feeding off a bird can take enough blood to cause problems. If this happens on a regular basis, then the sufferers will start to become anaemic and, if no help is forthcoming, may eventually die.

Tackling red mite is no easy task as these resilient little creatures reproduce at an alarming rate, and are difficult to get at with the various spray and powder treatments available for dealing with them. These mites mature and are able to breed just seven days after hatching, so treatment has to be applied every five days or so to stand any chance of breaking the life cycle. Spray-on treatments really need to be brought into direct contact with the mites to stand any chance of success, so care is needed with the application to ensure that the product is jetted into all the nooks and crannies where the mites may be sheltering. Partial dismantling of the house may be required to gain effective access: removal of the nest box, perches and roof panel. Houses made from tongue and groove-style timber planking can pose a real challenge, as every joint is a potential refuge for these tiny pests.

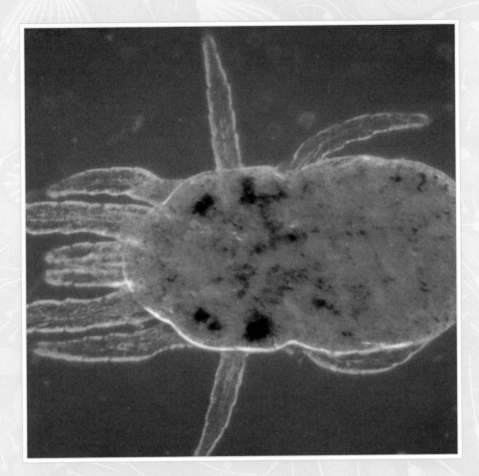

MITEY PROBLEM

Red mite colonise the nooks and crannies of a hen house (especially wooden structures with felt-covered roof panels). They come out at night, crawling along the perches, up the legs of roosting birds, through the feathers and down to the skin, where they bite and feed on the bird's blood. Once they have had their fill, they leave the bird and return to the security of their refuges within the structure of the house, to digest their meal, breed and rest during daylight hours.

The red mite may look innocuous but, when multiplied by thousands, its blood-sucking feeding habits can be life-threatening for chickens.

I've found grey, ash-like powder on the perches

CAUSE

This is a clear indication of the presence of red mite inside the hen house; action must be taken as soon as possible to put things right.

SOLUTION

There are two basic problems associated with the treatment of red mite. For a start, most of the anti-mite products currently on the market, and intended for use by enthusiasts at the hobby end of the chicken-keeping scale, rely on physical contact with the mite for their effectiveness. This is all well and good when you are able to catch mites out in the open, but is next to useless against those which are hidden from view when you are applying the product.

The other real snag is that an increasing number of the anti-mite treatments available nowadays are relatively ineffective. If you're unable to source creosote yourself – or find anyone able to apply it for you – then you'll have to do what you can with today's commercially available products. Seek advice from poultry specialist suppliers about what's most effective (new options are being introduced all the time) and, once you've settled on the product, be sure to follow the manufacturer's instructions to the letter, and make repeat applications. None of these products should ever be regarded as a one-off treatment: regular and thorough applications will be needed to stand any chance of keeping red mite under control.

This is typical of the grey, ash-like powder that you will find in areas where red mite are in operation.

TRADITIONAL APPROACH

Creosote is by far the most effective means of treating a red mite infestation. However, due to health and safety concerns its use has been restricted to professionals requiring its properties as part of their regular working practices. If you know one of these professionals, persuade them to spend an hour or so giving your hen house the once-over. Done properly, it will clear the red mite and enhance the longevity of your hen house.

43 My hen has patches of bare skin

CAUSE

Assuming the patches have not been caused by pecking damage from another bird, the most likely cause is the irritating and uncomfortable activity of parasitic lice.

SOLUTION

Lice are one of the most common parasites to be found on domestic chickens, and they feed by gnawing away at the base of the feather shafts and the bird's skin. Lice have brownish rounded bodies that can measure up to 4mm in length. They are easily spread from hen to hen within the flock, and are also common in the wild bird population, which is often how they arrive in your poultry pen in the first place. Their chewing can result in uncomfortable patches of raw skin, and the spread of disease. The female can lay several hundred tiny white eggs at a time, which she 'cements' securely on to the base of the feather shafts. These incubate and hatch within five days, producing nymphs that quickly grow to become fully functioning adult lice.

Spotting the presence of lice is the key thing; normally active chickens that rapidly start to appear bedraggled and lethargic should set the alarm bells ringing. Feather and weight loss, together with a reduction in laying performance, are other important pointers not to be missed. Check suspected birds carefully by hand, parting the feathers down to the skin and looking for both lice and eggs (adult lice can move surprisingly quickly as they dash for cover in nearby feathers).

Treat the affected birds with a good quality anti-louse powder, dusting thoroughly under the wings and around the vent, ensuring that the powder is worked right down to the skin. If the infestation is very bad, consult your vet and get a prescription for a commercial treatment such as Ivermectin, administered by injection or dropper.

CONSTANT IRRITATION

Out in the real world, among chickens that are allowed to range freely in a garden situation, it's very unlikely that they will ever be completely free from external parasites. As a responsible keeper, the best you can hope for is to keep creepy-crawly numbers in check with regular and effective anti-parasite treatment.

A careful, hands-on inspection is essential if you want to confirm the presence of parasites on your hens.

44 My chickens' legs have crusty deposits on them

CAUSE

The build-up of greyish-white deposits causing a general thickening of a chicken's legs is the sign that the bird is suffering from a scaly leg mite infestation. This is a tiny mite that makes its home under the normally tight-fitting scales of the leg, lifting them in an unpleasant and very uncomfortable way as it does so.

SOLUTION

There are a good number of decent treatments available, designed by specialist producers and intended to deal with the problem in a way that causes as little stress and discomfort to the chicken as possible. These treatments can also be followed up with a decent smear of petroleum jelly, which effectively acts to starve the mites of oxygen.

Another mistake that too many keepers make is to remove the lifted scales in the belief that this will – in some way – speed the treatment and healing process. Once again, this outdated practice does nothing apart from cause considerable pain for the bird. In fact, there is no good reason to remove any of the scales. Those that are damaged must be left until they are shed naturally by the bird, and replaced by new ones as part of the moulting process.

By following the treatment manufacturer's instructions carefully, you should see a fairly rapid eradication of the mites, but the subsequent healing process will be a slow one. Care and attention will need to be ongoing to ensure that any wounds clear up properly without becoming infected, and that those birds suffering don't become the pecking targets of other flock members. For this reason, and because scaly leg mite is contagious, sufferers are best isolated, and all housing should be thoroughly cleaned and treated appropriately.

BAD ADVICE

For years it was thought acceptable to dip the affected legs in surgical spirit to kill the mites. The effects of this liquid on the open wounds caused by the mite are agonising for the bird. Unfortunately, you will still see this appalling practice recommended in even recently published chicken-keeping reference books, whose authors should know better. The technique is also championed widely on the internet.

Leg scales should be tight and smooth on a healthy chicken. If you spot scales lifting and notice crusty deposits around them, the scaly leg mite is at work, and the bird will require immediate treatment.

45 My hens are gradually losing weight

CAUSE

The most common cause of a gradual weight loss among otherwise healthy chickens is the presence of intestinal worms, a very common internal parasite in domestic poultry.

SOLUTION

Intestinal worms can be introduced to your chickens when they eat earthworms, which carry their eggs, and via wild birds, which can also transport the eggs into the chicken run via their droppings. As a consequence, hens that are given access to your garden are never likely to be completely free from some degree of internal worm burden.

Fortunately, a low-level presence of worms in the gut does not pose a problem for a healthy chicken. It is only when worm numbers start to build that problems can arise. Environment plays a crucial role, too. Overstocked or heavily grazed runs are likely to cause trouble and the risks will be higher. The level of worm eggs will continue to rise in permanent runs as the background number is continually being added to by both the resident chickens and wild birds. Where a fixed run is the only option, regular top-dressing with fresh wood chip or sand layers can help minimise the problem. But the best option is to move the run on to fresh ground on a regular basis, reducing the risk of exposure.

The only way to assess the worm burden on your hens is to have a faecal egg count performed by a veterinary laboratory. This is not expensive to have done and provides an accurate indication of the real situation regarding internal parasites. Effective anti-worm treatments can be given as additives with either the food or drinking water. The recommended approach is only to treat when an actual problem is identified.

If you spot a worm in a chicken's dropping, it is a sure sign that there is a problem to be addressed.

GAPE, TAPE OR ROUND?

There are a number of worms that can cause trouble for chickens: the most common are the intestinal roundworm, the caecal worm, the hairworm, the trachea or gapeworm and the tapeworm. Of these, it is the roundworm and the tapeworm that are likely to be the most debilitating, and responsible for the loss of weight and reduced performance.

46 My hens are gasping for breath

CAUSE

While chickens can start gasping when being handled and stressed, birds seen doing it in a non-stressful, everyday situation could be suffering from the presence of an invasive little creature called the gapeworm, which typically lives in the bird's windpipe.

SOLUTION

The adult gapeworm is a small Y-shaped creature that infests the windpipe and bronchial tubes of young chickens (up to two months old). It can cause significant complications and, if numbers are allowed to grow, may even result in suffocation of the bird. The worms lay eggs, which are expelled by the affected bird, and these then develop over a seven-week period, with some even hatching out. The partially developed eggs (or the worms that have hatched from them), are then eaten by other hens, or by earthworms, which act as intermediate hosts and are often eaten by the hens. Once inside the new host, the young worms migrate to the lungs, then move up into the trachea, where they mature after about ten days.

Gapeworms cause severe inflammation in the trachea and the accumulation of mucus, both of which contribute to a restriction of airflow. As a result, the bird will be seen stretching its neck and 'gaping', open-mouthed, in an attempt to make breathing easier. Other symptoms can include headshaking, sneezing, gasping and gurgling noises.

Treatment needs to be swift and effective, using a suitable de-worming product, and a veterinary consultation may well be needed. Wormers can be administered in food, drinking water or via injection. If your birds are affected in this way, you should also think about keeping the youngsters inside until they are at least eight weeks old, and also relocating the run on to fresh ground. Isolating your chickens from the wild bird population, if possible, will also help as this is a ready source of potential re-infection.

WORMING THEIR WAY IN

While there are a number of different types of parasitic worm that can affect chickens, none of them should pose a serious problem if nipped in the bud. While routine, six-monthly worming is now thought inadvisable, keepers must remain ever vigilant and ready to act swiftly when problems are detected.

Chickens that are gasping for breath and frequently stretching their necks could well be afflicted with an unpleasant internal parasite called the gapeworm.

47 One of my hens is producing diarrhoea

CAUSE

Diarrhoea in chickens has many and varied causes: some serious, others not so. Regular observation of droppings is an essential part of good chicken-keeping practice, in order to identify problems early on.

SOLUTION

Worms are one of the causes of loose droppings, as are viral infections, bacterial imbalance in the gut, problems with the feed or diseases such as coccidiosis. Gut damage caused by worms can trigger disruption of the normal balance between good and bad bacteria in the bird's intestine, leading to an overgrowth of the bad types and the production of diarrhoea. Similar results can be prompted by a sudden change in food type, the consumption of dirty drinking water containing harmful bacteria, or the eating of mouldy feed that is tainted with fungal-type toxins capable of damaging the gut. If this sort of gut damage is allowed to continue and becomes excessive, it can lead to blood poisoning and potentially death. Providing your birds with a feed that is too high in protein can cause problems. This tends to trigger an increase in urate production, causing the birds to drink more which, in turn, moistens the droppings and can give the impression of diarrhoea.

Whatever the cause, diarrhoea is a debilitating condition that, if left unchecked, can promote lethargy, weight loss, dehydration and eventual death. Dealing with the problem, assuming there is no blood in the diarrhoea (which can indicate the presence of coccidiosis), should be relatively straightforward. Ensure that your birds have been wormed, are eating the correct diet and are given an appropriate tonic in clean, fresh drinking water. There are a variety of 'pick-me-up' type products available for chickens nowadays; consult your vet, specialist supplier or fellow keepers for the best advice about which to use.

BEAUTIFULLY FORMED?

Normal, healthy droppings should be firm, well-formed and show distinct brown and white sections. The latter is made from urates (the chicken's urine); chickens defecate and urinate in one motion. Caecal droppings, which are produced every nine or so motions, are caramel/toffee-coloured, sticky and smell very strong. These are completely normal and simply originate from a different part of the bird's intestines.

If the feathers around a hen's vent are messy like this, then all is not well. Diarrhoea is a typical contributor to this sort of condition.

48 My feather-crested birds keep scratching their heads

CAUSE

Crest mites are a problem that can affect any chicken breed with a feathery crest. While it is most usually associated with the Poland, it is something that Silkie, Houdan, Sultan and Araucana owners need to be aware of, too.

SOLUTION

Unless you are regularly checking the condition of your crested breeds by hand (which is highly recommended), the first indication of a problem will be an unusual amount of head scratching. This is prompted by the irritation caused by these mites; it can be so infuriating for the sufferers that they inadvertently damage their own eyes with their frantic clawing. As well as the head scratching, affected birds will become depressed and pale-faced, and egg production is very likely to suffer.

Careful treatment with a good quality anti-louse powder will help to put things right. This should be applied directly on to the head, ensuring that the powder is worked right down to the base of the feathers and the skin. Also, it might be advisable to use a cotton bud to apply a small amount of the powder gently into the bird's ear canals on each side of its head; crest mite may well be using this area as a refuge. Reapplication about a week later is very important, to catch any mites that may have hatched since the first treatment. Reinfestation is always a risk, though, so be on your guard. If you find it difficult to clear the problem, then talk to your vet about alternative, longer-lasting treatments that they may be able to prescribe.

HEADING FOR TROUBLE

The crest mite is a little brown creature that spends its entire life on the host bird. It looks quite similar to a red mite and is a very fast worker; it completes its life cycle in under seven days. This means that infestation is a typically rapid business and serious problems can develop very quickly. Clumps of crest mites that are left untreated on the bird can have a devastating effect, causing anaemia (they feed on the host's blood) and, in the worst cases, even death.

Crested birds, like this spectacular Houdan, need to be checked frequently for the presence of crest mites.

49 My chickens have lost their appetite but are very thirsty

CAUSE

These symptoms can be triggered by the presence of yet another common and potentially very troublesome external parasite, called the northern fowl mite.

SOLUTION

Northern fowl mite can be severe enough to cause anaemia (due to blood loss), heightened stress levels, a general loss of condition and a noticeable reduction in egg numbers. Keepers handling northern fowl mite-infested birds may notice the little pest crawling up their own arms, although only if numbers on the birds are really high. At other times, these tiny pests can be tricky to spot without the aid of a bright light and a magnifying glass. The best places to check include around the vent and under the wings, where you may spot the telltale signs of darkened, greasy-looking feathers that have been soiled by the mites' activities. Other key signs to watch for include a marked loss of appetite combined with an increase in thirst.

A keen eye for this pest is vital; its full reproductive cycle can be completed in under seven days in ideal circumstances, so numbers can multiply quickly. What's more, it has been estimated that a single host can be infested with as many as 20,000 mites. Fortunately, the likelihood of this sort of infestation becomes less likely once the bird is more than five months old. Sufferers should be washed carefully using warm water and a mild shampoo, thoroughly rinsed, gently dried and then treated all over with a pyrethrum-based anti-mite powder. If problems persist, seek a longer-lasting, spot-on type (a drop applied directly on to the skin, usually on the back of the neck) treatment from your vet.

A WINTER PEST

Like many other external parasites, the northern fowl mite feeds on the blood of its host. It lives on the bird at all times. Rather unusually, its most active periods tend to coincide with lower temperatures. Unlike other parasites, which tend to be most active during warm weather, the northern fowl mite does most of its destructive work during the autumn, winter and early spring months.

If you notice that your hens are drinking a lot more than usual but that their appetites have fallen away, then you could have a northern fowl mite outbreak on your hands.

50 My hen is pulling out her own neck feathers

CAUSE

This can happen as a consequence of quite a rare external parasite that is related to the scaly leg mite, but that works on feathers rather than scales, known as the depluming mite.

SOLUTION

While the scaly leg mite burrows its way under the scales on a chicken's leg, the depluming mite eats its way into the base of the feather shafts, close to where they emerge from the body. However, it doesn't do this all over the body, preferring to focus on areas such as the head, neck, back and upper legs. As a consequence of the burrowing, the chicken's body produces a nutrient-rich fluid, on which this particular mite feeds. It is painful for the bird, and the discomfort caused prompts it to scratch at and pluck out its own feathers in an attempt to get rid of the cause. This subsequent feather damage and loss is what gives the mite its name. Stress levels rise dramatically in birds fighting this sort of attack, causing them to lose weight, lay fewer eggs and become more susceptible to other disease- and infection-related problems. The breeding cycle of this mite – at about two and a half weeks – is not as fast as some of the others but, unusually, this creature gives birth to live young instead of laying eggs.

Successful treatment for this sort of infestation is not as straightforward as it is with the others, because the mites live in burrows, making it difficult to attack them with conventional, powder-type medication. The only real option is to use an Ivermectin-based drop-on product that will get into the bird's system and, thus, reach the parasites from the inside. Such treatments, however, are only available on prescription from your vet. What's more, their use will require an egg withdrawal period (no eggs should be eaten for at least a week after treatment), which must never be ignored.

The use of veterinary-prescribed, spot-on treatments represents one of the most effective and reliable methods of treating chickens for all manner of parasite-related problems.

STILL EFFECTIVE?

It is quite common for chicken keepers – even those with lots of experience – to inadvertently continue using medicine for their birds that is well past its 'best before' date. While this is unlikely to cause any harm, it won't do any good either, so always make a point of checking the label to guarantee effectiveness.

CHAPTER SIX

HEALTH ISSUES

Chickens are remarkable creatures, not least because of their ability to get on with life. Provide them with the health and welfare basics, and you can rest assured that they will reward you with trouble-free ownership, hours of interest and, in most cases, a wonderful supply of healthy, fresh eggs. However, despite a surprising degree of innate resilience, chickens can be sensitive souls, too. They are susceptible to stress which, in turn, lowers their resistance to infection, disease and the detrimental effects of internal or external parasites and any underlying medical conditions they may have.

Making sure that your birds remain healthy is, in many ways, tantamount to ensuring their happiness. Allow them the space to exhibit natural, active behaviour, keep them well fed and watered and provide them with a secure environment in which they can feel safe, and you will be giving your hens an almost perfect lifestyle in which they can prosper.

Observing your hens should be part of your daily husbandry routine. This chapter covers what to do if you notice worrying symptoms and behaviour such as excessive feather loss, swollen vent, lethargy or runny noses. Recognising the signs of health problems and acting swiftly and appropriately is key – no bird should be left to suffer unnecessarily.

51 One of my hens has a swollen, hard lower neck

CAUSE

The most likely reason for this sort of swelling is a condition known as crop impaction, a serious and potentially life-threatening problem.

SOLUTION

An impacted crop should be fairly easy to see and even more straightforward to feel while handling a bird. You may also notice weight loss in the affected bird. First thing in the morning as the birds are let out after a food-free night, there should be no sign of crop swelling; if there is, it is a cause for concern.

The fact that chickens have no teeth means that they rely completely on the mechanical action of the muscular gizzard (and the grit it contains) to break down what they have eaten, once it has been softened in the crop. Unfortunately, the system is not foolproof, and birds that eat unsuitable things like long, tough grass, feathers, wood shavings and straw can run into problems. These indigestible materials can become a tangled, unyielding mass in the crop that can be moved neither forward nor back. The resulting blockage can prevent the passage of food into the bird's digestive system, leading to weight loss and predominantly white, liquid droppings (due to the lack of solids being digested).

Birds suffering in this way will require treatment. You can dose them with bicarbonate of soda (one tablespoonful per 4.5 litres), or a small amount of olive oil (5ml). Combined with some gentle massage of the crop, this may be enough to soften and break down the tangle inside so that it can move along. If this doesn't work, the only option will be to seek a surgical solution. The blockage will need to be removed by a vet, via a small hole cut in through the neck and crop wall.

NORMAL SWELLING

When functioning normally, the crop – the pouch-like organ at the base of the neck – will fill as the bird feeds, then empty gradually as this food is softened and meted out to the gizzard (for grinding) and then on to the stomach. Birds that have recently eaten will show a swelling at the base of the neck, indicating a full crop, which is perfectly normal. Once that food has moved on, the neckline will return to normal.

Careful administration of a small quantity of olive oil, followed by some gentle massage, can help to soften and shift the contents of an impacted crop.

52

My chickens are hunched over and lethargic

CAUSE

If any of your chickens appear generally lethargic, with their feathers slightly fluffed up, and there is evidence of watery droppings and a reduced appetite, one of your first thoughts should be that they could be suffering the debilitating effects of coccidiosis.

SOLUTION

Coccidiosis is a tricky problem to tackle once it has been identified within a flock, because the infected birds are constantly shedding coccidia oocysts – essentially parasite eggs – in their droppings. These sit dormant on the ground, simply waiting to be picked up by other birds. They can survive for years in this state, well protected against extremes of temperature and even some disinfectants by their shell. Development is only triggered once an oocyst finds itself inside another bird, where the gastric juices in the gut break down the protective layer, releasing the infective form of the organism to begin its development. Resistance to this condition is achieved naturally in most cases, as young birds grow and are exposed to background levels of the parasite.

This is all well and good, assuming that the parasitic burden never gets too high. This is why it is important to keep your hens on fresh ground and not to overcrowd them. Hen houses and rearing sheds need to be thoroughly cleaned between batches of birds, and treated with specialist disinfectant products. As always, prevention is better than cure. However, take care not to go to the other extreme, as young birds need some exposure to the parasite in order to develop a natural resistance. Veterinary advice should always be taken if you are in any doubt. Anti-coccidial medication is available to treat badly affected birds, and specially treated chicken feed can be given to help prevent the problem in the first place.

If any of your chickens are looking obviously down at heel, you can be sure that they are suffering; immediate investigation is advised.

KICK IN THE GUTS

Coccidiosis is caused by parasites in the chicken's gut. The damage caused to the gut lining can have a serious impact on the bird's ability to absorb nutrients, so weight loss among sufferers is inevitable, together with diarrhoea. In really bad cases, the gut wall can be damaged to the extent that it begins to bleed, resulting in bloody droppings, anaemia and sometimes even blood poisoning due to bacterial infection.

53 Some of my chicks can no longer walk

CAUSE

This distressing symptom can be brought on as a result of chickens contracting Marek's disease, a virulent, herpes-like viral infection.

SOLUTION

Marek's disease is an unpleasant affliction, the consequences of which can be particularly distressing for inexperienced chicken keepers who are not used to dealing with suffering birds. It has several possible manifestations, which can present an assortment of neurological symptoms, including various degrees of paralysis of the legs and wings (temporary or long-lasting), tumours in vital organs such as the heart, ovaries and lungs, and skin-related issues, often leading to tumours in the feather follicles. Unfortunately, fatality rates are typically high, and infected birds will also be far more susceptible to other problems due to reduced resistance. In addition to the paralysis symptoms, which are impossible to miss, other signs of the disease include loss of weight, greying of the iris in the eye and raised or roughened skin around the base of the feathers.

Regrettably, there is no treatment once birds have become infected with Marek's disease, so culling is the only humane option. Vaccines are used by commercial hatcheries (on day-old chicks), but such treatments simply aren't economic at the levels at which pure-breed enthusiasts operate. Back-garden enthusiasts buying hybrid layers can usually be confident that their birds will have been treated but, while some domestic breeders of specialist breeds like the Silkie will implement a vaccination programme for their birds, others won't, simply because of the cost. So it's always worth asking the question before you buy.

INFECTIOUS SURVIVOR

Marek's disease is highly contagious, and is usually spread from bird to bird by the breathing in of infective feather-follicle dander. Furthermore, the virus can survive for a number of years at ambient temperatures in things like feather dust and, to make matters worse, is also resistant to some disinfectants.

Commercial hatcheries vaccinate day-old chicks against Marek's disease.

54 My chickens have become ill after new additions to the flock

CAUSE

When you add new stock to your flock, there is the risk that the new birds will transmit disease or parasites to the existing birds, or vice versa.

SOLUTION

Given that there is no pedigree system involved in the breeding of chickens, as there is with most other types of rare breed livestock, it can be all but impossible to find out exactly what you are buying and its state of health. Buyers – especially those with limited experience – are very much in the hands of the seller. What the seller decides to reveal about their stock's parentage and health status may be accurate, or it may not. The golden rule is never to take chances.

Every new chicken you buy should be placed in an isolation run for two or three weeks so that it (and its droppings) can be closely observed for its general condition and signs of illness and parasites. You also need to be sure about the new arrival's stage of vaccination. Most hybrids will have been treated against all the common poultry diseases at an early age, often using a live vaccine. Mixing these birds with your flock can result in any unvaccinated pure-breed birds becoming infected and deteriorating rapidly. Likewise, if your existing hybrid layers have been vaccinated but the two or three pure breeds that you decide to bring in haven't, the chances are that the new ones will start suffering soon after their arrival.

If you are re-homing ex-commercial laying hens, remember that these birds, having spent their lives in cages and not been in direct contact with droppings, will not have developed any immunity to coccidiosis (see page 116), so they should not be mixed with other cocci-resistant birds.

55

My hen seems depressed and has a hot abdomen

CAUSE

Hens that appear listless and whose abdomens feel hot to the touch are most likely to be suffering with egg peritonitis.

SOLUTION

This unpleasant condition can be triggered by a rupture in the oviduct or gut wall, or by egg material entering the abdomen rather than following its normal route down the oviduct. The result is that a yolk or a partially formed egg can be 'laid' internally, within the body cavity, where it runs the risk of causing an infection. If this happens, then the bird's decline can be rapid and in many cases it will result in death. Hens suffering in this way will become listless and disinterested, and stand around with their heads drawn in to their shoulders and their tails held down. When they are encouraged to walk, they will usually adopt an oddly upright, waddling sort of gait, a bit like a penguin. As well as the abdomen feeling hot to the touch (due to the infection inside), the bird's comb may turn a purplish-red or blue colour.

Shocks and stress can cause eggs to be laid internally on a one-off basis and, usually, that material will be reabsorbed gradually by the body without further problem. However, damage to the oviduct is likely to result in birds that lay internally on a regular basis, at a rate that is just too great for the body to deal with naturally. Any bird thought to be suffering in this way will require veterinary attention Regrettably, in these cases, vets will often advise that the bird is put to sleep to end its suffering. However, others may suggest removal of the oviduct to cure the problem, or even the insertion of a hormonal implant to halt the laying process altogether. Be warned that careless handling of in-lay hens can be a contributory factor to this condition; avoid squeezing laying hens as you hold them.

56 My chickens have watery eyes and runny noses

CAUSE

These symptoms, when combined with poor laying and slow growth, are the classic signs of a poultry disease called mycoplasma.

SOLUTION

The telltale signs of mycoplasma include watery bubbles in the corners of the eyes and discharge from the nostrils, which typically become swollen as well. You may also notice fluffed-up feathers, sneezing, gurgling sounds, strained crowing from affected males and lowered egg production from hens. All birds that contract the disease will remain carriers for life. It can prove fatal for some, while others may recover completely. It can then reoccur in these birds, although the subsequent bouts tend not to be as severe as the first.

The fact that the disease remains within the flock can pose problems for any new, mycoplasma-free birds that keepers may add to their existing group. The newcomers tend to succumb within a couple of weeks, while the rest of the flock remains outwardly fit and healthy. Wild birds can carry and spread the disease, too, so free-range hens will always be at risk. The disease can be spread via contaminated bedding and get carried from pen to pen on shoes, clothing, feeders and drinkers, although the infective agent (carried in the eye and nostril discharge) does not survive out in the open for more than a few days.

Given that mycoplasma is likely to be present to some degree in most flocks, minimising stress levels and maintaining good husbandry routines are the best ways to prevent potentially damaging outbreaks. Adding freshly crushed garlic or apple cider vinegar (cloudy, not food grade) to the drinking water is a great way to help hens maintain a healthy immune system. Treatment for mycoplasma is available from a vet, although the success of this is rather dependent on catching the problem early.

COMMON PROBLEM

Mycoplasma is a common disease among pure breed domestic chickens. There are a number of different strains, so the effects can vary considerably. The most frequently encountered type causes respiratory disease and weakens the immune system to the extent that all sorts of secondary issues – underlying health-related weaknesses which, under normal circumstances, wouldn't be of any consequence – start causing trouble.

Close inspection of your birds' eyes and nostrils should be part of your regular husbandry routine. Any sign of watery discharge can point towards the presence of mycoplasma.

57

One of my hens has a swelling on its foot

CAUSE

The most likely cause of this is a condition called bumblefoot, which is the result of an infection in the foot caused by a break in the skin.

SOLUTION

The key is to identify the problem before it gets too bad. It is very important to carry out regular, hands-on inspections of all your chickens and, as part of these, to look closely at the condition of their feet. Bumblefoot needs to be nipped in the bud, for the sake of the sufferer, because it is something that usually only gets worse, more painful and increasingly disabling. Unfortunately, the first thing that many keepers notice is that one of their hens has become lame, but by this stage the infection is already likely to be reasonably large and painful.

It is certainly possible to lance the swelling yourself, using a sharp scalpel so that the infected matter can be squeezed out, but most novice keepers will prefer to entrust this sort of task to an experienced vet. Pus will need to be drained from the swollen area and, in bad cases, it is likely that a solid core will have formed at the centre of the swelling, which will have to be removed as well. The wound will then need to be washed carefully and treated with a suitable antiseptic or saline solution, after which the bird must be isolated and kept inside on clean, softwood shavings.

This treatment procedure may have to be repeated before all the infection is removed. If the hock joint (further up the leg) feels hot to the touch, then it is likely that the infection has spread, meaning that veterinary help will be required.

Chickens should have the underside of their feet checked on a regular basis. Impact damage or splinters that break the skin can result in a painful condition known as bumblefoot.

IMPACT DAMAGE

There are any number of ways a chicken can cut or damage its foot. Two of the most common are splinter damage from a badly finished perch or hardwood chippings on the run floor, or repeated impacts from jumping down off a perch that has been set too high for the breed in question. This is more of a danger with the large, heavy breeds, or with birds that have been allowed to become overweight.

58 My hen's vent is red and very swollen

CAUSE

This is most likely the result of a vent prolapse, caused when a section of the oviduct from inside the hen gets pushed out through the vent.

SOLUTION

The traditional view is that hens that suffer a vent prolapse will need to be culled because of the risk of reoccurrence. However, most chicken keepers take the view that with some careful attention and plenty of TLC, hens with this problem can be saved and go on to lead happy, productive lives thereafter. There will be a risk of future problems, though, so it is essential that keepers remain vigilant.

The successful rectification of a vent prolapse will largely depend on your experience and, to a lesser extent, your stomach for this sort of operation. As a first step, the affected hen must be isolated from the rest of the flock, otherwise it is likely to suffer further from pecking damage; hens will typically turn on a flockmate with any sort of bleed or exposed wound/flesh. The protruding section of oviduct will need to be very carefully cleaned and then assessed for obvious signs of pecking damage. If this is present, or the area has become very dirty, then a course of antibiotics may also be required to fight off any resultant infection.

Next, the oviduct will need to be manipulated back inside the vent by hand, gently and slowly. Some keepers opt to apply a haemorrhoid-type ointment at this stage, but you should consult your vet, or another chicken keeper who has practical experience of the technique. You should also seek advice about the best cleaning product to use, and how to change the hen's diet to halt the production of eggs while healing takes place. It is important to appreciate that procedures like this are not always successful, and that prevention – by rearing and feeding your hens correctly – is always better than cure.

PROLAPSE POINTERS

There are a number of reasons why a prolapse might occur. The most common are as a consequence of attempting to lay an egg that is simply too big, or when a bird comes into lay but is actually too young to do so. Other factors include excessive fat, a calcium deficiency in the diet and the irregular passing of large faecal masses that over-stretch the cloaca (the opening through which droppings and eggs are passed).

A healthy hen's vent should be clean and pink. Any sign of swelling or fleshy, raw-looking protuberances should be regarded with suspicion, especially on young hens just starting to lay. You could be looking at the beginnings of a nasty condition called a vent prolapse.

59 A breeding hen is losing the feathers on her back

CAUSE

This is quite a common occurrence during the breeding season, when the male's 'treading' action on the female causes her to lose feathers.

SOLUTION

Chicken mating is an essentially rough and rudimentary process. While it is typically over in a matter of seconds, male birds – being larger and stronger than the females – are very dominant during the act. They will jump on to the hen's back (referred to as 'treading'), use their strong claws to grip her back and sides for balance, and often steady themselves by grabbing a beakful of the hen's neck feathers, too.

This rough treatment is not usually a problem, but there are instances when this can become a potentially serious issue. If there are too many males within a breeding group, then the females are likely to be in for a rough time. It is usually best if pens are restricted to a ratio of one male for every three females during the breeding season. But even when numbers are carefully controlled, there can still be problems. Quite often a male will settle on a favourite hen, and this poor bird will receive far more attention than the rest.

The clawing and feather pulling associated with treading, when repeated over and over again, can lead to feather loss and even skin damage. Keepers need to be aware of this potential for 'over-worked' hens to get injured, and to step in before things go from bad to worse. The put-upon bird can either be removed from the pen altogether, or she can be fitted with a poultry saddle. Traditionally made of leather but now available in more comfortable and lighter canvas, the saddle is designed to cover the hen's back and flanks, to protect her from the male bird's sharp claws and spurs. Loops fit over the hen's wings to hold the saddle securely in place.

LICE WATCH

It is important to remove the poultry saddle every few days so that you can check underneath it for the presence of lice or mites; it will get warm and damp under there, making the ideal breeding ground for parasites. Hens wearing saddles should always be well dusted with good anti-louse powder beforehand, and this should be replenished at each inspection.

Fitting hens with a poultry saddle will protect them from being damaged by the claws of an over-amorous male. Remember, though, to remove the saddle every few days to check and dust the wearer for parasites.

60 My hen has a cut and bleeding comb

CAUSE

If a bird's comb is bleeding, it could be the result of aggressive behaviour from another bird or an accident.

SOLUTION

Combs and wattles are vulnerable to stray pecks from other birds, and to getting caught on brambles, sharp edges or wire fences. A bleeding comb – or any other part of the chicken, for that matter – requires swift action to save the victim further suffering. A bird with any sort of bleeding wound will quickly become the victim of attacks from other members of the flock and therefore needs to be isolated, preferably into a small hen house that is clean and freshly bedded with softwood shavings.

Next, it is obviously important to ensure that the wound is cleaned and disinfected, and that the bleeding stops. Combs and wattles can bleed quite profusely, so the use of a veterinary wound powder can be very effective. There are plenty of these on the market, and the best act as an antiseptic as well as drying the wound very efficiently. The bird must then be left (with a supply of fresh water and food) to convalesce quietly for a few days. It is essential that it is not reintroduced back into the flock until the wound is completely healed. The slightest sign of any bleeding or remaining infection will draw unwanted attention from the flock mates, and you will be back to square one in no time.

Obviously, this sort of treatment is only really suitable for minor cuts and scratches. Although chickens do seem to have a remarkable capacity to heal themselves, for any larger wounds that may require stitching you should consult a vet, just to be on the safe side. Always remember that you have a duty to minimise suffering; just because an injured hen is showing no outward signs of discomfort, the pain it is actually feeling could be considerable.

SEEING RED

Flowing blood is like a red rag to a bull for other chickens; they are drawn to it irresistibly. It does not matter how docile and friendly the group of hens usually is; if one of them gets injured and starts to bleed, they will peck at it madly. The situation can very quickly turn extremely nasty and it is certainly not uncommon for an injured bird to be pecked to death by its flock mates in just a matter of hours.

Combs and wattles are vulnerable to all sorts of damage. If they get cut, they can bleed quite heavily.

CHAPTER SEVEN
EGG PRODUCTION

Most new chicken keepers are attracted to the hobby by the prospect of a regular supply of delicious, healthy fresh eggs. While healthy young hens should each produce five or six eggs a week, they will only do so reliably if they are enjoying life in an unstressed, pleasant environment.

Sustaining good egg production is something of a balancing act; hens won't simply pop the eggs out like machines. There are plenty of factors that can upset their laying rhythm, from harassment by other pets or young children, to environmental pressures such as overcrowding or sub-standard husbandry.

Get the balance right, however, and your back-garden hens will reward your efforts with uncomplaining production of the most delicious eggs you will ever have tasted. Yolks will be more golden and richer than anything you can buy in the supermarket.

Not every egg that is produced is perfect: some defects are benign, but others could signal some sort of deficiency in the bird's diet, and it's important to know the difference. Chicken keepers also need to keep a close eye on laying behaviour, as eggs laid outside the nest boxes can easily be broken, leading to the undesirable habit of egg eating.

61 My hens have not started to lay

CAUSE

Often chickens won't begin laying as soon as you buy them because the moment at which they start is governed by their age. New keepers are completely dependent upon the honesty of sellers, with regard to the onset of laying.

SOLUTION

Hens typically reach the 'point of lay' – often abbreviated to POL – when they get to 24 weeks old. This is the time at which sexual maturity arrives and, with it, the ability to reproduce; hence the production of eggs. However, it's not an exact science, and the arrival of POL varies with breed, and even sometimes between different strains within a breed.

Problems arise when breeders, in a bid to save money, sell hens as being 'POL' when, in fact, they may be anything up to eight weeks too young. Outwardly, though, there's no telling the age of the hen, and the supplier simply blames the lack of eggs on the bird's stress, caused by the move to a new environment.

So, the answer is to source your POL hens from a reputable breeder who provides decent birds of the correct age. This way you'll avoid the irritation and disappointment of a sometimes lengthy wait for the first egg to arrive.

BEST BEFORE
JUN 20 11
SIZE LARGE

New hens may not
start producing eggs for a
month or two after you buy
them. It all depends on their
actual age.

62 My hens don't lay eggs every day

CAUSE

Contrary to popular belief, hens don't lay an egg every single day. Their bodies simply aren't equipped to do so, so you shouldn't be expecting it.

SOLUTION

It's important to manage your expectation with laying hens. Good though they are – especially the modern hybrids – they aren't machines and so variation in productivity is inevitable.

In basic terms, a female chicken's primary aim in life is to produce a clutch of eggs (typically about seven or eight) in a nest, and then to sit on them for three weeks until they hatch. It's the fact that we take the eggs away each day that keeps them laying. But their bodies aren't built for non-stop egg production and, as a rule, most hens will lay an egg a day for six days, and then have a rest for a day while their bodily systems recover. Each egg takes about 26 hours to produce, and there's always a short gap between one being laid and the next one starting.

For this reason, the laying time for each hen gets progressively later every day. Each hen is also 'pre-loaded' with a fixed number of eggs that it can lay during its lifetime, but whether or not it actually achieves this depends on many variables, primary among which are general health, quality of environment and welfare conditions.

Don't expect an egg a day from each of your hens; their bodies aren't designed to sustain constant egg production.

63 The eggs are tiny

CAUSE

At the start of its laying cycle, a hen eases itself into the egg-producing process with the production of a few tiny, often oddly shaped eggs.

SOLUTION

Popularly known as 'wind eggs', these little offerings herald the onset of laying proper, and so are nothing to be alarmed about. Often, if you crack one open, you'll find there's no yolk inside, just albumen (white).

In most cases, the bird will only produce a handful of these miniature eggs, before the size increases dramatically as she settles into her normal laying routine. Consequently, this phenomenon is usually restricted to young hens that have just reached point of lay. However, it may also occur in older birds, if they're subjected to some sort of shock, such as the alarm caused by a fox attack, for example. But, as with the young birds, it should quickly pass, assuming that all else is well. If it doesn't and the supply of undersized eggs continues from a mature laying hen, then there will be something else at the root of the problem and veterinary investigation will be required.

Tiny 'wind eggs' are produced by young hens right at the start of their first laying season. They're nothing unusual and the bird should only produce a few of them before getting into her laying stride with the full-sized version like these ones.

64 The eggs' shell colour isn't right

CAUSE

If your hens aren't producing eggs with shells of the anticipated colour, then it's most likely to be either a seasonal- or strain-related issue, rather than a specific medical problem.

SOLUTION

A number of the pure breeds (and, increasingly, some of the modern hybrid layers), are famed for the production of beautifully coloured eggs. For example, the Araucana lays wonderful green-shelled eggs, the Cream Legbar produces blue ones, while those from the French Marans can be dark, luxurious chocolate brown. Unfortunately, though, these desirable shell colours aren't guaranteed, and it's only generally the best strains of each breed that deliver the goods in this respect.

The prospect of such gorgeous shell colours can work to heighten demand for these breeds among hobby keepers and, regrettably, this is something that less scrupulous breeders trade on. Sellers can make all sorts of claims about likely eggshell colour, and can thereafter come up with 101 plausible-sounding reasons why the birds aren't actually delivering. The reality, of course, is that they are an inferior strain.

So, always buy specialist breeds from a breeder who has been recommended by the relevant breed club or society. It's also worth mentioning that shell colour is always better early in the laying season, and will tend to fade as the weeks pass and the bird's reserves of colouring pigment are consumed.

BUY WITH CONFIDENCE

Everyone loves unusually coloured eggshells but, unfortunately, it's not simply a case of buying the appropriate breed to guarantee the desired result. There are many strains within each breed, and you need the right one to ensure a decent egg colour. The best way to find these birds is to buy on recommendation, via the appropriate breed club or society.

Much of the show-related breeding that has gone on in recent decades has been at the expense of egg production and shell colour so, when sourcing birds from a breeder, or public sale like this one, be sure that you know what you are buying.

I keep finding broken eggs in the hen house

CAUSE

Eggs can get broken by clumsy hens or because their shells are substandard and weak, making them easier than they should be to crack.

SOLUTION

Finding broken eggs in your hen house is really bad news, not only because it's such a terrible waste, but also because it can promote the very undesirable and hard-to-break habit of egg eating. Once chickens get a taste for raw egg, following a few chance encounters with the content from broken ones, they'll make the link and start cracking into other, unbroken ones to get at the goodies within.

The first thing to do is establish what's causing the problem in the first place. Your success at dealing with the problem will be very dependent on swift, effective action. The most basic precaution you can take is to remove the eggs from the house as soon after they've been laid as possible. Sometimes, though, this isn't always possible if there's nobody at home during the day. In this case, you should investigate 'roll-away' nest boxes which, as the name suggests, result in the egg rolling out of harm's way the moment it is laid.

If the shells are weak, which greatly increases the risk of breakage, then check the birds' diet; calcium deficiency is a common cause of this. Also, some feather-legged breeds are more inclined to clumsiness and egg breakage, but clipping their leg feathers can make a useful difference.

TEACHING BAD HABITS

If egg eating has become a regular occurrence, then you'll need to identify the culprit(s) and remove them from the hen house. The risk is that if you leave one bird that does it in place, it will teach the others to do the same.

Broken eggs in the hen house, whatever the cause, are bad news, and should be removed as soon as possible.

66 The eggshells are too dirty

CAUSE

Nobody likes the look of dirty eggs, and they're even more off-putting when you realise that what's on the shells is typically faecal matter. This can be picked up from soiled feathers around the hen's vent, or from contaminated bedding material onto which the egg is laid.

SOLUTION

There are obvious health risk implications associated with dirty eggs, both for those consuming them and potentially for the developing embryo inside if they are to be incubated. Either way, they're bad news and represent a problem that needs to be dealt with properly.

Hens with dirty feathers around their vents will be suffering from some form of health issue: a healthy chicken will produce predominantly dry, well-formed droppings with distinct white and dark brown/black portions. The sort of watery, diarrhoea-like droppings that will get caught in nearby feathers can be indicative of a number of conditions, including coccidiosis and viral infections, both of which will require veterinary attention. Dietary problems can be the cause, too, as can the consumption of mouldy feed or dirty drinking water.

A more obvious cause of trouble can be the presence of bedding material that has been contaminated by droppings in the nest boxes. There really is no excuse for this; it highlights poor husbandry standards and a generally low level of care and attention towards the birds. The easiest way to avoid this sort of problem is to 'poo pick' on a daily basis from the nest boxes. This is a simple job and need only take a few extra seconds when you let the birds out every morning. Not only will doing this solve the dirty egg problem, it will also greatly enhance the useful life of the bedding material in the nest boxes.

SMART SHELLS

While it is acceptable simply to brush off the worst of any shell surface contamination before storing eggs for domestic consumption, those intended for hatching should be washed carefully using a proprietary egg disinfectant and lukewarm water. Never use hot water – it can encourage potentially harmful impurities to be drawn inside, through the porous shell.

Dirty shells are bad news, not only from a visual point of view but also if you're intending to hatch them.

67 Eggs are being laid on the hen house floor

CAUSE

This usually happens because there are insufficient nest boxes for the number of hens using the house, or because those nest boxes are inadequate in some respect.

SOLUTION

Eggs stand the least chance of getting damaged if they are laid in a nest box. Any laid on the main house floor can be accidentally smashed as birds jump down off perches, or cracked when they get inadvertently kicked around as the birds enter or leave the house.

To encourage the chickens to use the nest boxes, they need to be well away from the pop hole and any windows, freshly layered with a good thickness of clean bedding material and, ideally, set above the main floor but below the level of the perches. The size of the nest boxes can be an issue, too: if they're too small, the birds won't feel comfortable using them. Chickens tend to all lay at round about the same time of day, so the nest boxes can be in demand. So, if there aren't enough of them for the number of hens using the house, then this can encourage floor laying. The rule of thumb is to provide one nest box for every three birds. Lastly, young birds may have to be encouraged to start laying in nest boxes. This can be done by placing one or two false eggs (made from china or plastic and often referred to as 'pot' eggs) in the nest box.

Ambient light is a key factor too, with hens preferring darker rather than lighter conditions; hence the need to position the nest boxes appropriately within the house. If this is wrong within the house that you have bought, it is a fundamental problem which is beyond complete rectification. The best you can do is to nail some hessian curtains loosely across the entrance to the nest boxes to cut down the amount of light inside, whilst still allowing the hens easy access in and out.

In order to minimize the risk of breakages, don't leave floor-laid eggs like this hanging around longer than necessary.

BAD BEHAVIOUR?

Problems like floor-laid eggs, bird-to-bird bullying or feather pecking tend to result from environmental failings. Get the basics right, though – the correct number of birds using a well-designed house and living a stress-free lifestyle – and you will find keeping a group of hens in the back garden an absolute breeze.

68 I am not getting normal eggs from my hens

CAUSE

Eggs can be laid with various shell defects: everything from coarseness and ridges to extreme thinness or even no shell at all. This can be down to poor diet, disease, stress and hereditary factors, amongst other causes.

SOLUTION

While much depends on the breeding stock from which the layers came and, indeed, the way in which the birds producing the eggs are being kept, perhaps the most closely linked factor is diet. A typical eggshell contains just over 2g of calcium carbonate, and this accounts for some 95 per cent of its overall weight. Interestingly, you will also find magnesium and phosphorous, together with tiny amounts of potassium, sodium, zinc, manganese, iron and copper. If a bird is missing out on any of these important constituents, it can result in sub-standard shell quality.

Soft-shelled eggs (where the egg is contained within a rubbery membrane) can be laid as a result of disease, fright, management changes, hot weather, hereditary factors or a vitamin D deficiency. However, it is not unusual to get a few soft-shelled eggs from pullets as they first come into lay, especially if this happens to coincide with rapidly increasing day lengths. Shell quality deteriorates each year towards the end of lay, as the shell-forming gland becomes worn out. This can also be quite a common occurrence with modern hybrids at the end of their laying season.

So, if you want the best eggs, you need to make sure that your hens are happy and healthy and enjoying a good quality balanced diet. If you find soft shells being produced out of the blue, then make some oyster shell grit available to the birds; simply put a few handfuls in a container near to the main feeder, and they will find and use it as necessary. Normally, though, this should not be necessary as well-formulated poultry feed should contain ample calcium to ensure good shell quality.

THIN BUT STRONG

Shell quality is a key aspect of a laying hen's performance, both from a practical and an aesthetic point of view. Eggs should look right to be appetising, but their shells also need to be strong enough so that they can be handled normally. It is really irritating, when reaching to pick up an egg, for one of your fingers to punch straight through the shell.

Eggs with shell defects can be laid as a result of all manner of triggers, but well-bred, healthy, relaxed and happy hens shouldn't produce anything like this.

69 My pure-breed hens are not delivering eggs

CAUSE

The most likely reasons for not getting a decent supply of eggs from pure-breed chickens are that you've got a poor strain, that they are overweight or that they have simply got too old.

SOLUTION

Many of our most popular pure breeds have been 'developed' by fanciers specifically for exhibition. Typically, this process involves selecting for far more feathers than the original breed ever had and, quite simply, more feathers equal fewer eggs. In general terms, chickens can produce one or the other effectively, which explains why they stop laying during the annual moult.

So, if eggs are important to you, you need to make sure that the pure-breed birds you're buying are from a recognised utility strain, not an exhibition one. Unfortunately, these have been lost completely in the case of many pure breeds, but can still be found if you pick a dual-purpose breed such as the Sussex or Dorking. Food intake is another important issue and, ideally, should be regulated to about 125g of good quality, formulated feed per bird, per day. Feed too much and your pure breeds will get fat, which, as well as posing serious health risks, will impact on their laying performance.

Finally, expect laying performance to fall away with age. Birds over four years old will lay appreciably less than those in their first flush of youth.

Breeds like this Orpington, which have become exhibition favourites around the world, have undergone generations of careful selection by breeders keen to develop the profuse feathering. Success with this has hit egg production hard, so this is not a breed to keep if you're after birds that lay well.

CHAPTER EIGHT

INCUBATION

Hatching eggs is one of the most fascinating aspects of chicken keeping. Of course, using an incubator is not what Mother Nature intended but, for most people breeding birds in the back garden environment, it is the most practical option. However, what a broody hen makes look deceptively simple is a delicate balancing act when eggs are placed in the unnatural environment of an incubator. Factors such as heat, humidity and egg turning – which the mother hen handles instinctively – are fundamental to the success of the whole process. Incubators need to maintain a consistently accurate control on these basic yet essential requirements, and their ability to do so varies with the sophistication of the machine and the way in which it is used.

Whatever the salesman or the marketing blurb may lead you to believe, never forget that there is a definite art to incubating eggs reliably. There is plenty that can go wrong and a great many inter-related factors that need to be juggled with skill and understanding. Storing your hatching eggs correctly, checking for fertility, and knowing whether or not to assist with hatching are just a few of the issues to be considered. But do not be put off: experience will, in time, lead to success, and the end result will be hugely rewarding.

70 I'm confused about how to store hatching eggs

CAUSE

There is a lot of conflicting information about how best to store fertilised hatching eggs before incubation actually starts.

SOLUTION

There are two golden rules that you need to bear in mind when storing hatching eggs before incubation. The first is that you should never, on any account, keep them in the fridge. The germinal disc inside the egg – from which the embryo develops – is sensitive to extremes in temperature, and there is certainly a risk of it being killed off by the sort of near-freezing temperatures found in most domestic fridges these days. Ideally, eggs intended for hatching should be kept at a constant temperature of about 12°C, and certainly out of direct sunlight, which is likely to cause dramatic temperature fluctuations inside. Temperatures that are too high can trigger the start of the growth process.

The second vital factor to bear in mind relates to time. Once an egg gets to more than a week old, hatchability rates start to fall away at a rate of 1.5 per cent a day. The rate of this decline increases beyond the 14-day mark, so any eggs that are much older than this are not generally going to be worth incubating. The ideal point at which to begin incubation appears to be when eggs are three days old.

Some breeders advocate storing eggs on their side, while others prefer having them either pointed end up or down. By and large, this appears to be a matter of personal preference, but the important things are to keep them separated (to avoid accidental damage and cracking) and to ensure that they are turned (ideally end to end) on a daily basis. Regular turning during storage periods is important to avoid the risk of the yolk becoming fixed in an off-centre position within the egg, which may prove detrimental to embryo development further down the line.

SENSIBLE STORAGE

The most sensible thing to store your hatching eggs in is a cardboard egg tray made specifically for the job (you can use normal egg boxes from the supermarket, too), which will guarantee their safe storage – and make them easy to turn – while they await the start of incubation.

Store hatching eggs carefully before incubation begins. Keep them secure, avoid direct sunlight and never put them in the fridge.

71

I do not know if my hatching eggs are fertile

CAUSE

Chicken keepers are often unsure of whether their hatching eggs are fertile, because there are no outward signs of fertility.

SOLUTION

It is very unlikely that every egg you set in an incubator will result in a successfully hatched chick, and there is no way of telling whether an egg is fertile or not from the ouset. The best you can do is to work on probability, based on your own experience and (hopefully) first-hand knowledge of the quality of the parents that produced the eggs.

Once the incubation process has begun, you can use the technique of candling to check for signs of development inside the egg. It involves nothing more complex than shining a bright light through the egg from behind, to highlight what is occurring inside. While you can use a torch and a cardboard tube for this, more consistent results will be achieved using a purpose-built egg candler. If all is well, then at the first candling session – after the eggs have been in the incubator for about ten days – it should be possible to see a spider-like shadow in the centre of the egg. The dark blob at the centre is the embryo, and the 'spider's legs' are the developing blood vessels. If there is nothing at all to be seen after 12 days of incubation, then the embryo hasn't developed for some reason, or the egg was infertile in the first place. These eggs are known as 'clears'.

Alternatively, development may start and then stop. This can be due to a number of things, including poor temperature control, incorrect humidity level, disease/contamination or because the egg was produced by poor quality parents. Any dud eggs should be removed immediately they are discovered; they can be a source of harmful bacteria that might affect others in the batch.

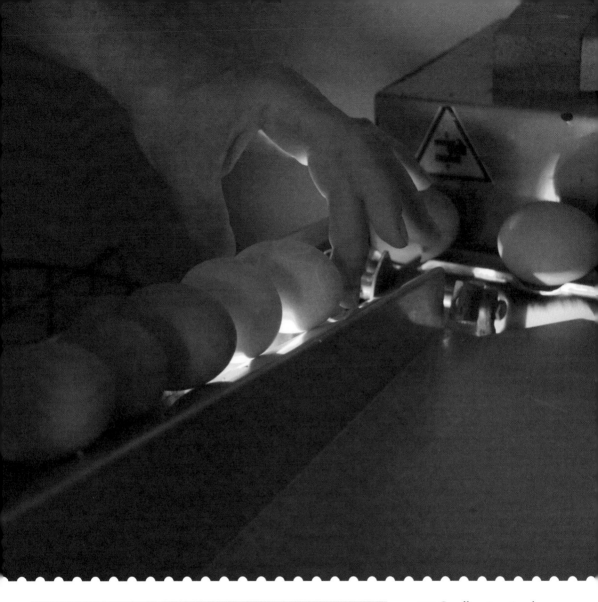

SAVE THE ENVIRONMENT

Avoid the temptation to candle the eggs too often, as each time you do so you will have to open the incubator, which, in turn, will have a potentially detrimental effect on the temperature and humidity settings inside the machine.

Candling is a simple technique that offers the only way of telling whether or not an egg is fertile. The first check should be made ten days after incubation begins.

72 I do not know whether to assist with hatching

CAUSE

Chicks can fail to hatch for all manner of reasons but, often, there is a fundamental problem that is stopping them from breaking out of their shell. Whether to help or not is a tricky dilemma that is usually best resolved with judgement from the head, not the heart.

SOLUTION

Anyone hatching eggs in an incubator will, sooner or later, observe one or more chicks that are simply unable to get out of their shells. They may make the initial hole – a process known as 'pipping' – using the 'egg tooth' on the end of their beak, but get no further than this. Alternatively, they might manage to struggle free of most but not all of the shell, only then to run out of energy. The big question, of course, is should you intervene?

There are two basic but opposing points of view on this matter. On the one hand, you have the survival-of-the-fittest attitude. Proponents of this view consider that if a chick is not equipped to get out of its own shell, then it will not be equipped to deal properly with life thereafter. While this may appear a callous, hard-nosed approach, their argument is that there is little point in breeding poor examples, and that allowing sub-standard birds to survive and be bred from in the future does nothing apart from weaken the overall gene pool and, ultimately, harm the breed.

On the other side of the fence are those who believe that life should be preserved whenever possible, by taking whatever measures necessary to ensure that the chick hatches and goes on to enjoy the best chance of a normal life. On the face of it, it is easy to sympathise with this compassionate point of view, but it is not a simple matter of life or death. The struggling chick may be ill and/or seriously deformed. Its quality of life and welfare thereafter is a vital factor, as is the obligation we all have not to prolong suffering of the birds in our care.

A DELICATE OPERATION

Assisted hatching is by no means an exact science. You could attempt to tweezer away pieces of shell, but however delicately you work there is a good chance that you will actually cause more damage to the bird. So, as always, think extremely carefully before you act.

To assist with hatching or not? It is an emotive subject and, for most breeders, a tricky decision. Head and heart tend to conflict on this one!

73 Not all the eggs have hatched

CAUSE

It is very rare that every egg in an incubator batch will hatch and, in reality, a failure rate of 10 to 50 per cent is to be expected. The actual figure will be breed-specific and dependent on the quality of the eggs you set and how your incubator operates (and is operated!).

SOLUTION

You cannot ensure that every single egg in a batch will hatch, but there are steps you can take to give as many of the eggs possible the best chance of hatching. As a general rule, it is best not to attempt to incubate eggs that are more than about a week old. The hatch rate starts to fall away dramatically after a week. By trying to hatch ageing eggs, you will be needlessly wasting incubator capacity, time and money. It is best to mark each egg with the date it is collected; write this carefully on the shell using a pencil (never a felt pen – shells are permeable).

When selecting eggs to incubate, avoid those that are badly soiled, as setting eggs with shells that are contaminated with faecal matter will promote the development of potentially harmful bacteria. You can buy specialised egg-washing products; some people use them, others do not bother. If you opt for washing, never submerge the eggs in water, and always use a recognised product. It is also important to allow the eggs time to dry thoroughly before transferring them to the incubator.

Finally, do not try to hatch eggs displaying any sort of shell defect or odd shape; always go for the ones that are the most classically egg-shaped. Those that are too round or too elongated should be avoided, as should any eggs with rough, patchy, ridged or thin shells. Anything that takes the egg's appearance away from the norm points towards an issue with the hen that laid it, which, in turn, raises serious question marks over the quality of what is inside.

A full hatch is quite a rarity and most breeders tend to expect a failure rate of 10–50 per cent, depending on the breed.

BALANCING ACT

As is so often the case with nature, the natural processes involved in a broody hen hatching an egg appear simple until you try reproducing the trick in an artificial environment. Getting an incubator to mimic the way a hen regulates temperature, humidity and egg movement consistently is difficult, as the cheapest machines often demonstrate!

74 I have noticed fluctuating incubator readings

CAUSE

Even slight variations in the temperature or humidity levels inside an incubation machine can have a pronounced effect on hatching success. Most commonly, fluctuations in readings are down to set-up or positioning errors, though the machines themselves can be inconsistent.

SOLUTION

One of the key fundamentals of correct incubator set-up is location, something that is frequently overlooked by inexperienced incubator users. While modern machines are designed to control the environment within their casings, their ability to do this not only assumes correct set-up, but sensible positioning, too. For best results, an incubator must be placed in a stable environment: meaning a room that is not too hot or cold, not too dry or damp and not subject to wild temperature fluctuations.

Many people – especially those with children who are keen to keep an eye on how things are developing – put their shiny new incubator in a hot kitchen, a heated utility room or a sun-drenched conservatory. All such options are deeply unsuitable. Centrally heated rooms, where the temperature is relatively high during the day and low at night, are not a good option either, and the fluctuations in humidity will put added strain on the system. Likewise, siting one of these units anywhere it is likely to receive direct sunshine at some point during the day is not advisable. The best locations tend to be well-ventilated spare bedrooms (with the central heating turned off), or brick-built garages shaded from the sun.

READ THE INSTRUCTIONS

All too often, chicken keepers who are excited to get started with their brand-new incubator fail to read the instructions properly. Given the delicate nature of the embryo development process the machine is trying to control, it is essential to take your time to read and understand the manufacturer's instructions. You must get the basics right to stand any chance of achieving decent hatching results.

While modern, electronically controlled incubators are certainly pretty good at maintaining ideal conditions for successful chick development, they are not miracle-workers. Operator error remains a factor, especially for inexperienced chicken breeders.

75 There is a yellow-red mass on the chick's tummy

CAUSE

Most commonly this is the result of an unabsorbed yolk sac, but it can also be triggered by a yolk sac infection.

SOLUTION

The yolk of an egg is absorbed into a chick's body through the navel as hatching approaches, providing a highly nutritious reserve that sustains the chick as it recovers from escaping its shell, until it starts to feed for itself. Problems arise when the yolk hasn't been fully drawn in before the navel closes. The remainder is left as a blob-like appendage, and is a sign of poor chick development. Once the chick hatches, it can become infected, which may prove fatal. This condition can be the result of poor incubation techniques and the use of hatching eggs from older hens, which produced eggs with yolks too large to be absorbed.

Yolk sac infection is another source of serious problems at this stage, and is typically caused by bacteria getting into the yolk before it is absorbed into the chick. Nasties such as E. coli can find their way on to (and then through) the porous shell via dirt and/or faecal contamination, or as a result of poor egg-washing technique before incubation starts. Another potential cause of trouble can be the 'explosion' of an infected but infertile egg inside the incubator, which can pepper the rest of the batch with contaminants. This is why it is always important to remove 'clear' eggs (eggs that have been identified as infertile) from the incubator at the earliest opportunity.

It is also possible for harmful bacteria to enter the chick's navel after hatching. Regrettably, most chicks infected in this way will not survive for more than a few days and, even if they do, they will be undersized, poor examples. As always, cleanliness is one of the best defences against this sort of problem. Incubators must be kept spotless and disinfected.

76

A chick has died in its shell

CAUSE

Known as 'dead in shell', this upsetting event has many possible causes.

SOLUTION

Death in the shell late on in the incubation process can be caused by the incubation of eggs that are too old, an incubator that has failed to maintain the correct temperature and humidity consistently, rough handling of the eggs before incubation, nutritional deficiencies or other shortfalls in the parent birds, a lack of effective egg-turning during the incubation process, a bacteria-contaminated incubator, or the presence of a 'lethal gene' within the genetic make-up of the chick.

As well as offering an ideal environment in which to hatch eggs, incubators provide perfect conditions (warm and damp) for growing bacteria. So, cleanliness is essential, and even brand-new machines should be carefully cleaned using a recognised sanitiser (your incubator supplier will be able to advise on the best brands). Cleaning should also be a top priority between egg batches.

Some of the more basic incubators come 'pre-set', in terms of operating temperature and humidity. These still need to be checked, so, before you start the process, it is important to run the machine for 24 hours, during which time you make regular checks on the temperature using an accurate thermometer (it should be held at a constant 37.8°C). Once you are happy with this, turn your attention to the humidity. This can be more tricky to assess, unless the incubator features a convenient humidity level display. Without this you will need to buy a humidity gauge. At the start of the incubation process, the humidity level needs to be between 40 and 50 per cent. Achieving this may require the addition of water; much depends on the season. The unit's operator manual should provide all the practical model-specific information you need.

77 My internet eggs have not hatched

CAUSE

When purchasing hatching eggs over the internet you are taking a gamble on two fronts: the quality of the eggs themselves may not be up to scratch and, even if they are, they may be damaged in transit.

SOLUTION

Buying hatching eggs 'blind' off the internet is a risky business. It is all too easy to be seduced by attractive-sounding deals. But it can turn out to be a false economy when, after putting considerable time and resources into the incubation process, none of the eggs that you bought at bargain prices hatch. The best approach is always to buy your eggs from a recommended, established breeder, and to collect them in person. If you do opt to order online for delivery, then you should do your research first and find a reputable supplier.

If the eggs themselves are in good order when dispatched, delivery can create its own problems. Often the handlers are none too gentle – even when packages are marked 'Fragile' – so the packaging must be of a high standard if the eggs are to survive the trip undamaged. Sellers who offer to pack their eggs in the specialist, polystyrene modules that insulate and hold each egg securely are likely to be the best bet. Even when the eggs are placed in such packaging, however, some couriers still manage to shake the life out of hatching eggs, and the resultant hatch rates from the batch can be disappointingly low. Allow eggs a day to settle after they have been delivered, then carefully check them for signs of shell damage. Use a candling light to inspect the condition of the air space, which you will find at the blunt end of the egg. This should be clearly visible and intact. If it is not, or appears to be full of bubbles, then it is likely that the egg has been shaken around too much and is not worth incubating.

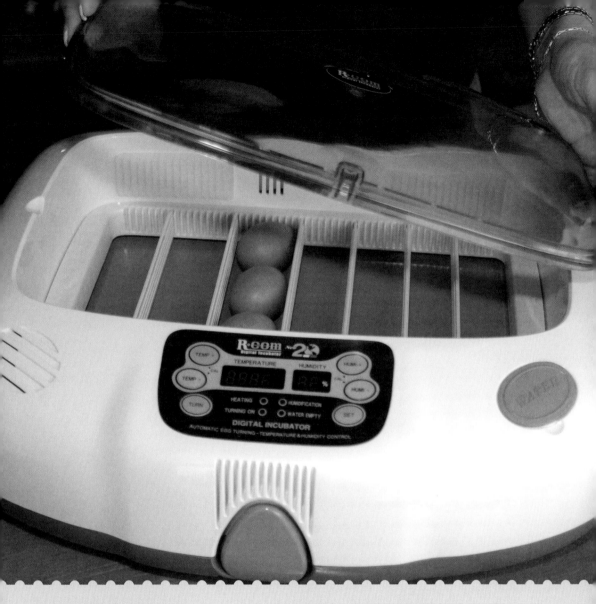

Hatching eggs that have been delivered through the post is always going to be a bit of a gamble.

MAKE THE EFFORT

Rather than placing your trust in the careful hands of a postal or courier service, take the time and trouble to collect hatching eggs yourself, straight from the breeder. Not only will you know exactly how the eggs have been handled, you may also pick up some wonderful nuggets of wisdom, free of charge!

78 The eggs are not hatching on time

CAUSE

Chicken eggs should take 21 days to hatch and, assuming all is well, it is a process you should be able to pretty much set your watch by. Chicks that take longer to emerge typically do so because there is a problem, either with the chick itself, or as a result of poor incubation conditions.

SOLUTION

The two most common causes of late hatching are setting eggs that are stale and running an incubator at a consistently low temperature, the latter simply slowing the whole development process. It is also worth noting that hybrids, in general, tend to hatch earlier than pure breeds, and that some of the latter are slower than others.

The success of the incubation process is inextricably linked to the quality of the hatching egg. If it is too old, has been mistreated, is damaged or appears non-standard in any way, then the chances of a healthy chick emerging from it after three weeks of incubation time will be dramatically reduced. The answer – to incubate only eggs that are as near 'perfect' as possible – appears a ridiculously obvious one, but it is a maxim that is ignored by far too many chicken keepers, who are subsequently disappointed with the hatch rate achieved. Of course, there can be mitigating circumstances. In cases where the number of hatching eggs is very low, for example, those wishing to incubate may be forced to go with what they have got.

While leaving a batch of eggs in the incubator for longer than the standard 21 days may result in a few more chicks hatching, it is likely that the quality of the ones that hatch late will be poor. As a general rule, the first out always tend to be the best. Chicks from a hatch that starts late but continues in the normal way should be OK, while hatches that start on time but then drag on should be more of a concern.

Any abnormality in the shell quality or overall egg shape should be enough to cause an egg to be rejected as an incubation prospect.

CHAPTER NINE
REARING

Chicken keepers should never underestimate the importance of the rearing stage in a young chicken's life. After all the excitement of hatching, the tendency can be for keepers to take their foot off the diligence pedal a little. In fact, this is exactly the point when close observation should start. Time spent simply watching your birds is never wasted, regardless of their age. Getting into the habit of doing this during the early growing stages is key: you can tell so much about how young chickens are developing simply by watching their behaviour. Careful observation also allows you to spot trouble at the earliest opportunity, giving you the greatest chance of nipping problems in the bud.

Chicks are vulnerable during their first few weeks of life, so it is vital that you do all you can to ensure that they have a safe, clean and comfortable environment in which to grow and develop. The type of brooder you opt for will therefore be an important decision. Just as with the incubation stage, temperature is a key factor in rearing chicks, and you will need to safeguard against potential brooder accidents such as crushing and drowning. There are health issues and diseases that are particularly relevant to chicks, such as Marek's disease and spraddled legs; as always, knowing the signs and acting swiftly is key.

79 The chicks are not interested in eating

CAUSE

While most chicks will happily start eating chick crumb as soon as they are able to, some will not. The reasons for this are varied and include poor incubator performance, bad ambient lighting and sub-standard parent stock.

SOLUTION

Chicks start out in life with an internal supply of food (the yolk from the egg is absorbed into the body shortly before the chick hatches), intended to last for a day or two until the youngster is strong enough to start eating on its own. However, problems can arise if, for whatever reason, a chick is unwilling to eat and drink. Youngsters exhibiting this kind of reluctance will quickly fall behind in their development compared to chicks that are eating normally, and you will need to get them interested in food sooner rather than later if you are to save them. If there is a physical problem with the chick, perhaps related to a defect that developed inside the egg, or some problem that has been inherited from sub-standard parents, then the prospects for recovery are not great.

Some traditional breeders believe that incubator-hatched chicks are at a disadvantage compared to those hatched under a hen, because they miss out on the instructional examples set by the mother. A little demonstration of how to peck at food can be all that a reluctant chick needs to start doing it for itself. One trick is to sprinkle chick crumb down on to a small piece of cardboard, so that it makes a pitter-patter sound within earshot of the chick. Alternatively, use the tip of your finger to make a pecking movement down on to the food on the card. The tapping sound that this makes can also work well as a trigger to encourage youngsters to start pecking for themselves.

EGGS FOR CHICKS?

Some traditionalists feed their newly hatched chicks on a finely chopped mixture of hard-boiled egg and chives. The youngsters find it very palatable and, with it being a vitamin- and mineral-packed option, it can provide the chicks with a very nutritious introduction to eating for themselves.

While most chicks will start eating and drinking perfectly naturally, some may not. Keen observation is required to make sure that all are behaving normally during the early stages.

80 I am not sure which type of brooder to get

CAUSE

There are a number of different brooder options available, ranging from the cheap and cheerful to the relatively expensive. Picking the right one can seem a bit daunting, until you appreciate the basics of brooding.

SOLUTION

You will need an appropriately sized, well-ventilated container in which to keep the youngsters safe and secure. There are no specific guidelines for brooder space allocation in terms of chick numbers, so use your common sense to avoid overcrowding, ensuring enough space for the youngsters to move around freely and feed and drink with ease. A perfectly effective brooder can be made from a sturdy cardboard box or, if you have some DIY skills, you could make your own out of wood. It is also possible to buy ready-made units, while some larger incubators also offer some degree of brooder functionality, although only for the first few days. In most cases, breeders use an open-topped, box-type container when the chicks are still very young. Wire mesh is used over the top to prevent the chicks escaping, as well as unwanted visitors getting in. Escapees will very quickly become chilled and, if left outside, they will surely die.

Another key requirement is maintaining the right temperature (at the start it must match that of the incubator from which the chicks have been moved), and usually this is done by employing some form of heater. Nowadays electricity provides the cleanest and safest power option. Infra-red heater bulbs are a popular choice, although some believe that the red light produced can overstimulate the youngsters. It is also possible to buy a ceramic-coated bulb called a 'dull emitter', which gives off heat, but no light. This allows the breeder to create a natural day/night light pattern, which is thought to be advantageous to chick growth and development. Doing this, of course, requires a separate white light source.

Brooding chicks require an appropriately sized container where they can be kept safe, warm, fed and watered.

AN ELECTRIC BROODY

You can buy something called an 'electric hen', which is essentially an electrically heated panel that is mounted on adjustable legs so that it sits a few centimetres above the floor of the brooder. It is supposed to mimic the presence of a mother hen, and the chicks can move in underneath it whenever they feel they need some warmth.

81 My chicks have messy bottoms

CAUSE

This condition can be caused by a poor brooder temperature, the youngsters getting chilled during transfer to the brooder, excessive nutrient intake or the use of unsuitable ingredients in the food.

SOLUTION

Pasted vent is a relatively common problem affecting chicks at this stage of their development. This, as you might have guessed, can be recognised by the presence of dry or drying faecal matter around a chick's vent. Often the vent will become capped, so will have to be carefully cleared. If the condition persists and you have done all you can to correct the likely causes, then you will require veterinary help.

It is paramount that you keep the brooder clean and dry. Wet and dirty brooders provide an ideal breeding ground for one of the main risk factors for chicks, coccidiosis, the symptoms of which can include diarrhoea. Fortunately, you can now buy effective and approved anti-coccidiosis products. Fresh drinking water and feed in containers (all frequently cleaned) that are easily accessible for the birds must also be provided, too. Brooder positioning is another aspect to be considered. Never position it in direct sunlight as temperature fluctuations need to be avoided. A brooder unit will always deliver the best results if placed in an airy (but not draughty) environment where the ambient temperature is reasonably stable, at somewhere between 12–18°C.

Finally, you need to be confident that your brooder set-up works consistently, and the only way to be sure about this is to test it first. Run it constantly for a day or two while taking regular temperature readings at just above floor level. You need to be confident that your heat source is capable of maintaining the correct temperature whatever the time of day or weather conditions outside.

Careful observation is a key requirement when rearing chicks. Keep an eye on general behaviour and watch for conditions such as 'pasted vent', something which points towards a poor brooder environment.

82 The chicks are all huddled under the heater

CAUSE

If you find your chicks bunched together under the heat source, it is an indication that the birds are cold.

SOLUTION

All chicks, from a day old, need a carefully regulated heat source in order to survive. Fortunately, there is a very simple way of checking whether or not chicks are happy and contented inside their brooder. All you have to do is look at the way they are spread around inside the unit. After they have had an hour or so to settle following the move from the incubator, observe the birds. If you find them huddled together directly under the heat source, then this is a clear sign that they are too cold, so the temperature will need to be checked and the heat source adjusted accordingly. Alternatively, finding them spread out around the perimeter of the brooder unit, away from the heat source, suggests that they are too hot. If you have the temperature just right you will see a well-spaced and constantly moving group of chicks that appear active and alert.

The overall temperature will need to be gradually reduced as time passes and the birds develop their first proper feathers. The starting point should be 37.5°C, matching the temperature of the incubator from which they are transferred to avoid a temperature shock. Measure the temperature at a point just above floor level, so that it reflects most accurately the conditions in which the birds are living. It should be kept constant for the first couple of days, then reduced by 3–5°C every week. This rate of reduction can continue for three to four weeks (or until the temperature reaches about 18°C). By that stage, the birds should have grown their first set of proper feathers, and can be taken 'off heat' altogether.

ADJUSTABLE HEAT SOURCE

The simplest way of controlling temperature reduction is to mount the heat source on a chain and gradually raise it above the brooder floor by a couple of links at a time. You will have to experiment to find out what works best for your set-up. Ideally, you should do this before the brooder is occupied, so that you are not working on a trial and error basis with the birds.

Chicks all huddling together directly beneath the heat lamp in a brooder like this will usually be doing so because they are cold.

83 Some of the chicks in my box brooder have died

CAUSE

It is always very distressing when chicks die, especially if the deaths could have been prevented. Most commonly, overcrowding in the brooder is to blame, resulting in individuals being pushed into corners and smothered to death.

SOLUTION

Chicks need space to be able to move around freely within the brooder, to feed, drink, warm up and cool down as and when they feel the need. Suffocation is an ever-present and deadly risk associated with overcrowding, and one of the most frequent causes of chick deaths in the brooder. Chicks have a tendency to gather together in corners and, if there are too many of them, the pressure on the poor souls trapped at the back can be overpowering. One neat way to alleviate this risk is to round off the corners of a square unit with curved, cardboard inserts. It is a simple but very effective modification.

The golden rule about avoiding overcrowding is one that you will encounter over and over again as your involvement with chicken keeping evolves. The trouble is, there aren't any hard-and-fast rules about the amount of floor area required by a set number of chicks in the domestic situation, so it largely comes down to a matter of common sense. Be observant at this important stage and keep a wary eye on the behaviour of your chicks. Overcrowding and incorrect temperature setting are common causes of trouble. If the chicks are cold they'll naturally huddle together, which can lead to injuries among the weaker or smaller specimens. Also, a heat lamp that's overheating will force the youngsters away from it, leading once again to the risk of crushing and smothering injuries or fatalities if the birds haven't got enough space to move to an appreciably cooler part of the brooder.

BROODER BASICS

Overcrowding can lead to all sorts of problems. Tightly packed chicks will be far more likely to begin pecking each other out of boredom. They will also compact and contaminate the floor litter far too quickly, causing ammonia levels to rise, which, in turn, greatly increases the likelihood of serious respiratory and eye-related problems. Wet and dirty litter can also harbour a deadly disease that affects chicks, called coccidiosis.

Rounding off the corners in a brooder box is a simple but very effective way to reduce the risk of chicks being smothered.

84 I found a drowned chick

CAUSE

Newly hatched chicks are relatively unsteady on their feet, making them extremely vulnerable to drowning if the wrong type of drinker is used.

SOLUTION

Newly hatched chicks have a greater need for drinking water than they do for food, and so it's vital that they have an easily accessible and plentiful water supply. However, it is also very important that the water supply does not pose any risk of drowning. Young birds are unsteady on their feet, and can easily be accidentally knocked into a drinking bowl by their brooder mates. If the water has any depth to it at all, they could very easily drown.

The sure-fire way to avoid this is to make use of a properly designed chick drinker, which is like a scaled-down version of a standard, gravity-fed chicken drinker. It presents the water in a shallow and narrow trough-like ring around the base, and the liquid is replenished as it is consumed by the birds. If for some reason you are unable to get hold of a proper chick drinker, and have to use some sort of bowl or dish instead, be sure to limit the depth of the water in it. The best way to do this is to add a layer of clean marbles or small stones to the base of the container. This will ensure that while water is still available to drink, there is no significant depth. However, you will need to keep a special eye on the consumption rate to ensure that the level never falls too low, making it difficult for the chicks to get at the water among the marbles or stones.

One other drinker-related point worth noting is that some breeders recommend placing the food at one end of the brooder unit and the drink at the other. This encourages activity between the two ends, which is believed to be good for healthy muscle development.

Newly hatched chicks need water more urgently than food to begin with, so make sure that it is provided in a clean and safe way.

85 A couple of chicks have done 'the splits'

CAUSE

This is a condition known as 'splayed legs' that affects newly hatched chicks; in most cases, it is reversible.

SOLUTION

Splayed legs is a relatively common condition among very young chicks, although there is some debate about what actually triggers it. One theory is that it is more common among chicks that have hatched in incubators where the humidity has been incorrectly set. Another suggests that it is more likely if chicks are left in the incubator for more than the recommended 24 hours' drying time after hatching. Many believe that the condition of the incubator floor plays a significant part. If it is too smooth, young birds can slip and easily dislocate one or both of their legs; with their weak muscles and still pliable bones, newly hatched chicks are particularly vulnerable. Breeders often cover incubator compartments with corrugated cardboard to minimise the risk of slippage.

The good news is that splayed legs can usually be put right with a bit of careful attention. Remedial treatment involves gently moving the legs back into their natural position and then loosely tying them together with string to stop the same thing happening again. After a day or two of being 'hobbled' like this, the chick will usually have regained enough muscle strength to walk normally again when the string is removed.

Splayed legs is just one of a number of physical conditions to watch out for with young chicks. The point at which you transfer them from the incubator to the brooder unit is the ideal time to give each one a quick inspection. The sort of things to look out for include twisted toes, twisted beaks and missing toenails. Once the youngsters are in the brooder, watch for individuals that do not seem to be displaying normal active and inquisitive behaviour like the rest.

FOR THE GOOD OF THE BREED

If you are hatching for breed conservation reasons, youngsters showing any defects should be removed, both for the good of the breed and the long-term quality of your own stock. It is always important that only the best, strongest and healthiest chicks are reared.

When you transfer chicks from the incubator to the brooder unit, it's a good opportunity to give them a quick inspection, looking out for things like twisted toes and beaks and missing toenails.

86 Small chicks are being picked on by the larger ones

CAUSE

Differences will occur between chick sizes if batches of different breeds or ages of birds are mixed together in the same brooder. Growth rates can also be affected by eating habits and underlying health issues.

SOLUTION

As with so many other aspects of good chicken keeping, the more organised you are with your breeding routines, the better your results will be. So, when it comes to hatching and rearing, there really is no substitute for a methodical approach. Consistency is always the key, so the eggs you set in your incubator should ideally all be from the same breed and more or less the same age.

Likewise, the batches of chicks moved from the incubator to the brooder should be similarly categorised; populating a brooder unit with a hotchpotch of youngsters of different ages is just asking for trouble. Even at this early stage in their lives, the young chickens will be all too well aware of differences, and so individuals that stand out because they are an odd size (or unusual in some other physical way) will be much more likely to attract problems. In the same way, mixing large fowl with bantams can lead to a similar effect. By and large, it will be the bigger birds that are most likely to pick on the smaller, weaker ones. Feather and toe pecking are common consequences and, left unchecked, this sort of aggressive behaviour can quickly escalate into the far more serious problem of cannibalism.

It is always best to ensure that only birds of the same size and age are put together in the brooder unit.

87 The chicks are not feathering well

CAUSE

Slow feathering is just one of the problems that can be caused by poor brooder management, and in particular inadequate temperature control.

SOLUTION

When it comes to brooding young chicks, a common mistake that novice chicken keepers make is to keep the unit too warm, preferring to err on the side of too hot rather than risk the chicks becoming too cold. In reality, though, too much heat can be really bad news, often resulting in poor feather growth and/or feather pecking among the youngsters, plus suppressed appetites and generally slower than expected growth rates. For normal development to occur – including proper feather development – it is essential that chicks are kept at just the right temperature. Under natural conditions, chicks will regulate this themselves by staying out and being active until they start to feel chilly, at which point they return to the feathery warmth of their broody mother. Rearing chicks without the benefits of a broody hen means that the responsibility for providing the all-important heat falls on the keeper's shoulders.

The temperature inside the brooder unit must be accurately matched to that of the incubator when the chicks are first transferred, and then reduced gradually as the youngsters grow and develop. Keeping the temperature too high will remove the growth stimulus, slowing overall development. Conversely, a brooder that is too cold can be even more of a disaster, promoting chilling, which can lead to high mortality rates. It is also worth bearing in mind that exposing young chicks to sudden changes in temperature can provide the stimulus for disease.

88 My chicks have suddenly become ill

CAUSE

The vulnerability of newly hatched chicks means that the chances of them suffering from health problems are high if your brooder unit and husbandry techniques are not up to scratch.

SOLUTION

To keep your young birds fit and healthy in their brooder is not difficult providing you can offer them cleanliness, attention to detail, common sense and an appreciation of where the main threats to their health lie.

Mushy chick disease (also known as 'yolk sac infection'), is quite a common problem that is caused by bacterial contamination of the yolk sac in the newly hatched chick, typically resulting in death. It affects chicks in their first week or two of life, and most mortality will be seen during the first 72 hours after hatching. Affected chicks appear drowsy, droopy and listless; if they do move around it is typically with a staggering gait. They also present ruffled feathers and many that die do so because their weak condition leaves them unable to compete with healthy chicks for food and warmth. Dirty hatching eggshells or a contaminated incubator are the usual causes; the infective bacteria contaminate the embryo either through the shell before hatching or, once the chick is out, via its navel. In addition it can also be contracted orally due to unsanitary conditions in the brooder.

The disease can be treated by prescription drugs, although only with limited success, so veterinary advice will be required. If it is going to strike, rickets usually appears in chicks that are around four weeks old, and can be more common in some heavy breeds. The cause is related to vitamin D and calcium deficiencies, which promote lameness and even an inability to walk. Treatment is relatively simple via dietary supplements.

89 My chicks seem depressed

CAUSE

Chickens of all ages are extremely susceptible to stress induced by sudden change, so a badly managed transfer to a new environment is perfectly capable of sending them into a decline.

SOLUTION

By the time the chicks get to six or seven weeks old, their feathers will have grown and you will be able to turn off the heater. They will then need to be transferred to a larger, unheated coop. Ideally, this should be combined with a covered run, giving them access to the outside but without full exposure to the elements. It is vital that they are given plenty of time to become acclimatised to their new surroundings, and to continue their growth and development without stress-related interruption.

As in the brooder, food and water should be made constantly available, although, by this stage, the birds should be big enough to be switched from chick crumb to a good-quality grower's pellet (by gradually mixing in an increasing amount of pellets over three or four days). Continue with softwood shavings as a bedding material. The weather, of course, has an important part to play. However, assuming your birds have reached this stage in the spring, they will require a couple more weeks of restricted access before being given the freedom to range freely. This will also give them plenty of time to get used to the new food ration; combining a feed change with a move outdoors would represent a major, and unnecessary, trauma. Take care not to mix them with any other chickens you may have as they will still be vulnerable to infection. They should go on to fresh, clean short grass, well away from other stock.

WATCH WHAT THEY EAT

Do not put chicks on long grass that has been recently cut and is covered in the clippings as these can prove indigestible for young birds — causing potentially fatal compaction in the crop or gizzard.

Never be in a rush to get young birds outside, however much you may want to see them pecking and scratching around in the sunshine. All environmental transitions need to be gradual.

90 I can't tell the sex of my chicks

CAUSE

In most breeds, the characteristic sexual differences between male and female chickens, such as feather shape, tail type and comb size, don't become apparent for weeks, sometimes months, after hatching.

SOLUTION

The problem of not knowing the sex of chicks can be an expensive one for those keen on breeding, as it means that all youngsters have to be fed, watered and cared for until they are old enough to start displaying the telltale differences that will then allow accurate selections to be made as required. By and large, most breeders choose to keep many more egg-laying females than breeding males, so needlessly rearing surplus male birds can be a time-consuming nuisance.

The sex of a chick can only be determined in the so-called 'auto-sexing' breeds, such as the Cambar. Created from a series of breeding crosses made between the barred Plymouth Rock and a gold Campine, the Cambar produces chicks the sex of which can be identified from the moment they hatch; a light-coloured patch of down on the back of the head clearly identifies the females. A series of auto-sexing breeds – Legbar, Rhodebar, Dorbar and Welbar – was subsequently created, although the idea never really took off in commercial terms. A variation on the theme can be achieved by crossing a 'gold' male, such as a Rhode Island Red, with a 'silver' female (for example, a Light Sussex hen). This will produce gold-coloured female chicks and much lighter, lemon-yellow males. Also, mating barred males with non-barred females will give non-barred female chicks and barred males; these are identifiable by a spot of white at the back of the head of the day-old male. However, never forget that sex-linkage can only be relied upon if the parent stock are pure.

MALE OR FEMALE?

Mating any light-breed male with any heavy-breed female should enable the sex in the offspring to identified by the feather length in the wings at one day old, the female chicks showing more development than the males.

For those keepers breeding a lot of birds, being able to tell the sex of chicks straight out of the shell is a great asset.

CHAPTER TEN

BEHAVIOURAL PROBLEMS

While keeping hens certainly isn't rocket science, it is something that demands routine care combined with attention to detail. Your typical group of egg-layers won't be a demanding bunch, but their few requirements must be met if all is to remain sweetness and light. One of the secrets of good chicken keeping is developing the ability to recognise the signs of trouble, so that you are able to react and nip things in the bud before they get too serious.

As your keeping experience grows, you'll learn to recognise some of the telltale signs that all is not well. Part of this involves getting to know your birds' normal behaviour patterns, so that you are able to spot when things change. Chickens are creatures of habit, so variations in behaviour tend to occur for a reason. Develop the knack of spotting these deviations, and identifying their most likely causes, and you'll be well on the way to becoming a very accomplished chicken keeper.

A certain amount of bickering to establish the pecking order can be expected in any flock, but should fighting become a too regular occurrence you may need to step in and identify the troublemakers. And what to do when a hen goes broody and refuses to leave the nest box, or your chickens start eating their own eggs? All such undesirable behaviour is examined in this chapter.

91 I can't catch my chickens

CAUSE

There are two primary reasons for not being able to catch chickens: poor technique on the part of the catcher, and the fact that the birds aren't used to being handled. Obviously, the two are interlinked!

SOLUTION

It is essential that you are able to catch, hold and inspect your chickens; it's something that should be done on a regular basis to assess their general health, search for parasites and gauge weight, and so on. However, breed temperament plays a key part in the fundamental handleability of any chicken. Light breeds are inherently more highly strung and excitable than the typically docile heavy breeds and hybrids. But even hybrids can become difficult to deal with if they're never handled.

So, the most important thing is to get your hens used to being handled by doing it regularly from day one. Adopting a sensible method of capture is essential, too. Don't chase them around the pen like a mad thing; stay calm, quiet and slow-moving at all times. The easiest approach is to pick birds up from inside the hen house, when they're calmly roosting at night. But if you need to catch one during the day, carefully shepherd it into a corner of the pen, slowly move closer, then, when you're near enough, make a decisive grab. Place one hand down on the bird's back and use the other to scoop it up and gather it towards your body. It's not advisable to grab chickens by the neck, wings or legs, as damage is relatively easy to do.

BEING INTERACTIVE

The more time and effort you put in to interacting with your chickens, the better things will be. Establishing their trust should be a core aim, and spending time in the pen is the best way to do this. Regular handling, too, is a must if you're to properly monitor their health and welfare.

Movements when catching chickens in daylight always need to be slow and deliberate. The object is to keep stress levels as low as possible, for all concerned.

92 My cockerel is too aggressive

CAUSE

Feistiness or aggressive behaviour among male birds can be a strain-related character trait, but also often tends to be season dependent.

SOLUTION

Male chickens are naturally more aggressive than females; that's just the way it should be and is the natural order of things. How you deal with the problem rather depends on your situation, your attitude towards the offending bird and the reason why you're keeping it in the first place. Experienced breeders know that a good, fertile male bird will turn into an aggressive character during the mating season. This behaviour is expected and appreciated as it tends to denote a successful breeding bird, all else being equal.

This aggression rarely manifests itself as violence towards other birds, and is usually directed towards the keeper, or anyone else who happens to wander within pecking or clawing distance! This sort of behaviour can pose problems for those keeping their birds in a family environment, with young children who enjoy regular contact with the birds. Previously mild-mannered males can turn into feisty and unpredictably aggressive characters when they reach sexual maturity, typically at about six months old, so this is something to be very aware of, especially if you're breeding any of the larger breeds. Male-to-male aggression can be a serious issue, too, so breeders will need the space available to separate these birds once they mature. A naturally aggressive male bird that exhibits undesirable behaviour all the time is no fun to keep.

FEISTY MALES

It's not all bad news if you have an aggressive cockerel. Although he may be awkward to manage at times, the flip side of this feistiness is that more often than not he'll be an excellent breeding bird: virile and fertile. So, if you have ambitions to breed, a pugnacious male can be a great asset.

Male chickens, by their very nature, are more aggressive than females; they're noisier, too.

93 One of my hens has been attacked by the others

CAUSE

Chickens are drawn to the sight of blood; even the most mild-mannered hen can become a frenzied attacker under certain circumstances. What is more, attackers will show no mercy...

SOLUTION

The odd scuffle – the sort of thing that happens as birds establish the pecking order within their group – is nothing to worry about; it is a perfectly normal part of day-to-day life in the chicken pen. However, things become more serious if there is blood involved. A bird that gets injured badly enough to break the skin is likely to be set upon by the rest of its flock mates once they catch sight of the blood. A kind of bloodlust develops which, if ignored, can quickly escalate. Consequently, it is vital that keepers remain vigilant at all times, inspecting birds on a daily basis to check for signs of injury and/or bleeding.

Any injured birds will need to be removed from the main pen and isolated for their own safety. They should be kept apart from the rest until their wounds have been treated and/or healed naturally (depending on their severity). It is also worth noting that bird-to-bird pecking can be prompted by overcrowding, boredom resulting from a poor environment, and inadequate food. A sub-standard feed ration that is badly formulated or short on protein can prompt the birds to seek sustenance elsewhere, which they may do by pecking at the feathers (which are 80 per cent protein), first on the ground, then on flock mates. Always use a reputable brand of chicken feed and, if necessary, add a suitable dietary supplement to the drinking water (your vet or poultry specialist supplier will be able to advise). You can help alleviate boredom by suspending fresh vegetables in the run at head height for the birds, or by giving them pieces of fresh turf from elsewhere in the garden to peck and scratch at.

Keep a wary eye on the condition of birds with missing feathers; these individuals should be isolated at the first sign of blood, for their own safety.

PRIMITIVE BEHAVIOUR

Newcomers to the chicken-keeping hobby are frequently surprised – and often shocked – to discover just how vicious their feathery garden friends can be. Anything smaller than they are is regarded as a potential target of snacking on, and they won't hesitate to turn viciously on an injured flock mate, given the opportunity.

94 My hens keep pecking at each other's feathers

CAUSE

Feather pecking can be triggered by environment-based factors, such as a lack of space, a poor environment, enforced confinement, and, most difficult to deal with of all, straightforward bullying.

SOLUTION

To deal effectively with a feather-pecking problem you need to understand the common triggers for it. Overcrowding is one – something that all chicken keepers must avoid. Also, remember that hens need stimulation in their everyday lives. Those kept in cramped conditions or featureless runs will develop problems, both disease- and behaviour-related.

Bullying is a problem that can occur at any time, even if you have provided your chickens with the ideal conditions in which to live. Some birds are just more aggressive than others, and will almost inevitably cause problems for those lower down the pecking order. This sort of dominant behaviour often manifests itself in pecking attacks to the victim's head. The aggressor may stop the bird it is targeting from feeding or drinking, or from entering the hen house to lay or roost. Sometimes these situations sort themselves out as the bullied bird learns to keep out of the way (assuming there is sufficient space for it to do so). In other cases, though, it does not, and the situation simply worsens. When this happens, the only practical option is to remove the aggressor.

Pecking can also become a problem when new birds are introduced to an existing flock, which inevitably upsets the established order of things. Introducing newcomers at night, into the house when the flock is already roosting and settled, is generally regarded as being the best approach, but smooth relations are never guaranteed. While the birds can be allowed to re-establish the pecking order themselves to a degree, supervision will be required to ensure that the initial scuffles do not turn more serious.

Feather pecking can quickly get out of hand if the victim's skin gets broken and blood starts to flow. Keepers need to be vigilant at all times, so that problems can be nipped in the bud before they escalate to this level.

95 One of my chickens will not leave the nest box

CAUSE

The most likely reason for this is that she has gone broody. This is completely natural, instinctive behaviour.

SOLUTION

While obvious sitting and reluctance to leave the nest box will be a big clue that a hen has gone broody, added pointers can be that her feathers appear more fluffed up than usual, and that she is noticeably vocal and grumpy – possibly even aggressive – when you try to move her. Broodiness should not really be regarded as a problem, although plenty of keepers find it a nuisance. Dealing with broodiness really is a matter of personal preference.

Some keepers are happy to let it run its course, taking care all the while to manage the hen through her lonely vigil. Her inclination will be to sit constantly, so it is important for the keeper to lift her out of the house and encourage her to eat, drink and defecate – though you may have to take a peck or two for your trouble. The two or three weeks that it will last for represent a bit of a dead spell. Not only will the hen involved stop laying completely for the duration, she is also likely to lose body condition. Her weight will be reduced as her food consumption decreases.

Others prefer to try to 'break' the broody hen, most commonly by relocating her to a wire-floored compartment (with food and water, of course) that is raised off the ground. The idea of this is that the ventilation from below helps cool the hen's underside and, after two or three days of this, in most cases the broody inclination disappears. The hen can then be returned to its normal laying routine in the shortest possible time. Those birds that are allowed to sit, whether on eggs or not, will certainly benefit from a suitable drinker-based tonic to help recoup the lost body condition as quickly as possible.

BROODY BREEDS

Some breeds – such as the Dorking, Silkie and Sussex – are more highly regarded for their sitting reliability than others. Others are recognised as non-sitters, including many of the Mediterranean breeds like the Leghorn, Minorca and Spanish. Modern hybrid hens have had the broodiness bred out of them to maximise their laying potential, so you will very rarely find one that sits at all, or at least not reliably.

A broody hen will be determined to sit in the nest. Efforts to remove her will generate angry clucking and, usually, a peck or two.

96 One hen is losing feathers all over its body

CAUSE

Assuming the chicken is generally healthy and living in a decent environment with good welfare standards, feather loss from all over the body is most likely to be caused by the annual moult.

SOLUTION

Chickens going through a full moult can take on a worryingly bedraggled appearance as their old feathers are lost and before the new ones grow to replace them. It usually happens in late summer so that there is still plenty of time for the new set to grow before the winter weather arrives. The moult is triggered by the shortening day-length, and takes anything from three weeks to three months to complete. While it is happening, egg laying generally ceases (although there are a few exceptions): because eggs and feathers are very high in protein, the bird is unable to produce both at the same time. Priority is given to the feathers, and the whole process provides the bird with a period of rest and recuperation. Eggs laid after the moult are of much better quality than those laid just before the break. After a short break, moulting birds should return to full feathering and production.

Feeding during the moult can be approached in one of two ways. You can 'force' the moult by withholding the layers' rations and feeding only grain, which removes the birds' urge to lay eggs, allowing their bodies to concentrate fully on losing and replacing the feathers. Oats, bran and 'middlings' (ground wheat) are ideal feeds, but not maize, as this will cause the birds to generate too much heat. Alternatively, you can feed as normal, but supplement with plenty of greens, together with rapeseed or sunflower oil, both of which are beneficial to the birds as a source of vitamins and minerals. The additional water contained in the greens will also help to keep the chickens cool and well hydrated.

WHY THE LOSS?

It is important to realise that moulting is not the only
trigger for feather loss. Other factors such as bullying, poor
environment, stress, disease, parasitic infestation, cockerel
treading, and extremes of light and heat can all lead to feather
loss to some degree. In most cases, it is possible to isolate the
cause using a process of deduction, but, if you are completely at
a loss, then call in a professional for some experienced guidance.

*Keepers witnessing
the annual moult for the
first time can be shocked
at how their beloved
birds start to look!*

97 My chickens are eating their own eggs

CAUSE

Chickens love the taste of fresh egg and, given the opportunity, will eat them without a second thought. Accidentally broken eggs, or those produced with soft shells, can prove irresistible.

SOLUTION

In most cases, this sort of behaviour starts completely by chance, with birds encountering a broken egg in the nest box or on the house floor. One or more of the birds that share the house may taste the goodies inside the broken egg and, realising how delicious it is, start actively looking for more. If, coincidentally, there is a problem with shell quality among the eggs being laid, then the chances are that they will not have to wait too long for their next fix. As things progress, the egg-eaters will start hanging around, waiting for opportunities to present themselves and then, given time, they will learn to start breaking into eggs with normal shells, which is when the problems really start. Realistically, if this habit becomes ingrained then it is almost impossible to break; the only solution will be to remove the offending birds from the flock.

The important thing is to prevent the problem developing in the first place. Eggs must be collected regularly and as soon after they have been laid as possible. The producing of floor-laid eggs needs to be discouraged; if it is happening, then the trouble may lie with the nest boxes. Ideally, these should be raised at least 45cm above the house floor, and face away from the light. The addition of a hessian curtain covering two-thirds of the entrance to darken the interior of each box can help, too. Size and number are important considerations as well. Each nest box should be a 30cm cube, and there should be one for every three hens using the house. Traditional anti-egg eating measures, such as filling a blown egg with mustard or chilli paste, can never be guaranteed to work.

Egg eating is a common chicken vice that is both destructive and habit forming. Once hens get the taste for fresh egg they'll actively start looking for it and, what's more, they'll teach others in the flock to do the same.

98 I have a hen that has started crowing

CAUSE

Hens sometimes adopt male characteristics due to hormone imbalance. Typically, though, it is only for a relatively short period of time.

SOLUTION

In most instances sex-reversal in chickens can be explained quite simply. If, for the sake of argument, a chicken was to undergo a 'gonadectomy' (removal of the ovary or testes), both sexes will become neutral in gender; the male becomes a 'capon' and the female, a 'poulard'. Outwardly, both birds appear the same, too, with long, male-type feathers, spurs and a shrunken comb. The capon will remain so, but the poulard may change again, reverting back to its female-type feathering, but also developing the comb and wattles normally seen in males. Most of this can be put down to a lack of hormones normally generated by the gonads. The male hormones are responsible for the comb and wattle development, and the female ones for the feathering.

In practice, apparent sex change in hens occurs following an event that causes the ovary to stop working. Having laid for a season, they start to crow and develop male characteristics following the moult. Typically, they continue to display this male-type behaviour throughout the following season (some even laying eggs) before reverting back to their original form for year three. It is thought that during the moult the ovary becomes inactive, failing to secrete enough female hormones to cause the new feathers to be of the female type. Poulard feathers result, almost identical to male ones. As the bird comes out of the moult and the ovary starts to become active again, the bird will begin functioning as a female and start laying eggs, even though she retains the outward appearance of a male bird. At the next moult the same may happen again, or she may revert back to being entirely female.

THE DEVIL'S WORK

In days gone by, crowing hens and laying cocks were regarded with much suspicion, and were thought to be the work of the devil. Some even believed the eggs produced would hatch into serpents!

In any group of hens there will always be a leader: a bird at the top of the pecking order. Sometimes, though, hens can even start displaying male characteristics, like crowing!

99 My chickens are just too feisty

CAUSE

Keeping male birds can inevitably create friction within a group of chickens, with age often proving a key factor. However, breed is also fundamental to the general behavioural characteristics you can expect.

SOLUTION

The way in which you deal with feisty chickens very much depends on what's causing the problem in the first place. At its most basic level, the behaviour of any chicken will be influenced by its breed (see box).

Males (of virtually any breed) can turn from mild-mannered cuties to scratching and flapping firebrands once sexual maturity arrives. Such tendencies are always less pronounced among females, but can be seen among some strains. Groups of male birds that are reared perfectly happily will need to be split up and segregated as they mature to avoid problems. This requires extra space, equipment, and organisation. The Game breeds, in particular, will require careful management and novice keepers need to think carefully before getting involved with them; even the females can become quite a handful once fully grown, and birds which keepers start to find intimidating and difficult to deal with are only ever going to get worse as the control exerted over them inevitably lessens. Problem birds – of either sex – will become all but impossible to deal with once their aggressive behaviour has become ingrained.

Keepers finding themselves in this position have three basic options. The first is to isolate the offender and do what you can with careful care and attention to improve its behaviour; commonly, though, it's unlikely that such birds will ever be able to be safely reintegrated into the main flock. The second option is to re-home the offending bird but this, obviously, is never a particularly easy thing to do. Finally, there is the humane slaughter alternative, which really is the last resort.

Game breeds, like this powerful Asil, are best avoided by inexperienced keepers.

CHARACTER TRAITS

The soft-feathered heavy breeds – such as the Dorking, Orpington and Sussex – can be relied upon to display calm, friendly and docile characters. Light breeds, including the Leghorn, Minorca and Spanish, tend to be more easily startled and, consequently, harder to handle. Game breeds are the most aggressive. Breeds such as the Asil, Malay and Shamo were honed for cock fighting, and much of that inherent aggression remains to this day.

100 I have found a hen with a bleeding bottom

CAUSE

The hen concerned has most likely been the victim of vent pecking, a problem that stems from the interest chickens have in all things red.

SOLUTION

In most cases, the trouble starts out of nothing more than simple curiosity. After a hen has laid an egg, her vent will often stay reddened and slightly swollen for an hour or two. This, if spotted by a flock mate, can prove to be an irresistible pecking target, and with the flesh there being soft and exposed, damage is all too easily done. If a bleed starts it is likely to be fairly bad which, of course, only adds to the attraction as far as the other birds are concerned. Unless the injured hen is able to keep well out of harm's way until the bleeding stops, she can be in serious trouble, even facing a painful and protracted death if help does not arrive.

Assistance, in this case, is best delivered in the form of an observant keeper who, spotting the blood around the vent area (stained feathers should be pretty easy to see), swiftly removes the hen from the enclosure and installs her in a quiet, comfortable isolation unit. A hen house well away from the other birds is best for this; it is not essential that it has a run attached but, if it does, so much the better. The important thing is that the bird has access to food and water and that it feels both safe and relaxed in its temporary new home. The wound will need to be assessed and treated accordingly. Specialist disinfectant sprays and wound powders are widely available for dealing with this sort of eventuality if the damage is not too severe.

Bigger wounds, however, will probably need to be stitched by a vet. It is essential that the bird remains in isolation until the wound has healed completely. There may also be concerns about the inevitable stretching caused by the egg-laying process reopening the wound, so you will need to take professional advice on this aspect, too.

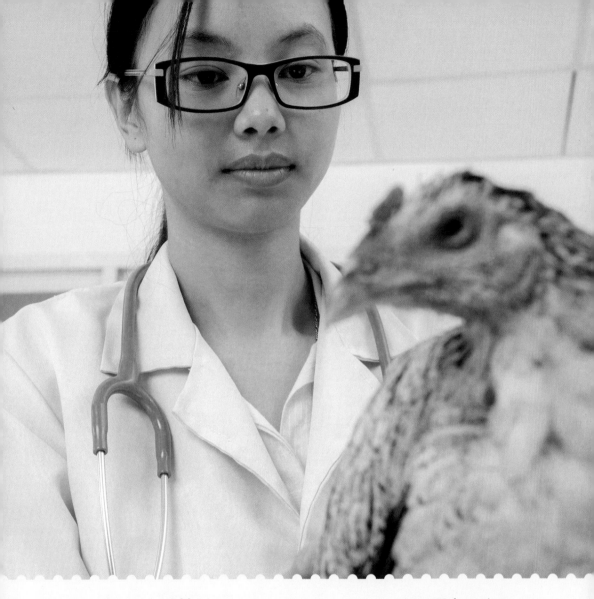

Unless you're an experienced keeper, a bird with any sort of flesh wound is best dealt with by a chicken-friendly vet.

GLOSSARY

AUTO-SEXING
A characteristic in certain breeds (e.g. Cream Legbar) whereby the colour of the chicks' down allows them to be identified as male or female from the moment they hatch.

BANTAM
A miniature chicken which should weigh one quarter that of its large fowl counterpart. Most breeds are available in bantam form nowadays.

BARRED
A plumage pattern distinguished by alternate, light and dark stripes running across the width of a feather.

BREAST
The front portion of a chicken's body: the area between the base of the neck and the point of the keel bone.

BROODER
A safe and secure enclosure into which chicks are moved once they've hatched in an incubator. Temperature must be matched to that of the incubator, then gradually decreased as the birds' feathers develop.

CANDLING
The use of a bright light to test for embryo development inside a fertile hatching egg. First candling usually takes place 10 days into the 21-day incubation process.

CAECAL DROPPING
Produced about every tenth time a chicken defecates. It is often yellowish and is much looser than the normal dark brown and white droppings.

CHICK CRUMB
Finely milled feed specially produced for consumption by young chicks, which are not yet able to cope with the standard-sized pellets fed to adult birds. Typically fed for the first six weeks or so, chick crumb can also be medicated to help control coccidiosis.

CLOACA
Also commonly referred to as the vent, the cloaca is the end of a chicken's digestive and reproductive tract, through which both the bird's droppings and, in the case of a hen, the eggs, are passed.

COCCIDIOSIS
Often abbreviated to 'cocci' (pronounced 'cocksi'), this is a painful, potentially serious disease caused by parasite damage to the gut wall.

COMB
The fleshy, typically bright red protuberance on the top of a chicken's head. Distinct types include single, rose, pea, walnut, leaf, horn and folded.

CREST
The collection or tuft of feathers on the top of the head of some pure breeds, such as the Poland.

CROP
A section of the oesophagus at the base of a chicken's neck, where food is temporarily held and the process of digestion begins.

DEAD IN SHELL (DIS)
A chick that dies in its shell at, or close to, hatching time. Potential causes include heat and humidity issues.

DOWN
The soft, fine, hair-like covering found on newly hatched chicks. Also describes the small, fluffy areas found at the base of many feathers.

EARLOBE
Folds of coloured skin that hang below a chicken's ears. These vary in size, shape and colour from breed to breed and can be an important indicator of breed quality.

FEATHER-LEGGED
Most breeds have clean, featherless legs but some, such as the Brahma, Cochin and Pekin, show a growth of feathers down the side of the leg and on to the toes.

FLUFF
The soft, down-like feathering found around the thighs on soft-feathered breeds such as the Cochin. Undesirable on most light and hard-feathered breeds.

GIZZARD
Compartment in the digestive tract with thick, muscular walls used to grind down food (typically with the help of grit), once it has been softened in the bird's stomach.

GROUND COLOUR
The main feather colouring of a chicken, on which the markings (such as lacing) will be overlaid.

HACKLES
The long feathers found on the neck and saddle (rear end of the back), most obviously seen on male birds, where they are typically pointed.

HARD-FEATHERED
The sort of close, tight feathering found on Game birds, such as the Old English Game.

INCUBATOR
A machine used for hatching eggs, in which temperature, humidity and egg movement are carefully controlled to ensure healthy embryo and chick development inside the shell.

KEEL
The chicken's breast bone, or sternum, running from the base of the neck to between the legs.

MOULT
The annual replacement of tired, old feathers at the end of the season with new ones. Typically occurs in late summer/early autumn, and takes several weeks to complete. Feathers are replaced in a set order, not all at once.

OFF-HEAT
The point at which young chickens are sufficiently developed and feathered to be able to survive without the need for artificial heat.

ONDULINE
A bituminous, corrugated material often used as an effective and durable roofing solution on hen houses.

PASTED VENT

Potentially fatal condition most commonly affecting young chicks; dried droppings build up in the feathery down around the vent, causing a blockage.

POINT OF LAY (POL)

The moment at which a hen begins egg production. Typically occurs at 24 weeks old, but can vary.

PRIMARIES

The main flight feathers of the wing. They are the longest found there, and the ones that need clipping to prevent flight.

PULLET

A female chicken that's yet to go through her first moult.

QUILL

The hollow stem of a feather, which attaches it to the bird's body. It also provides the support for the web of the feather; the flat, blade-like portion.

SADDLE

The rear portion of the back, reaching as far as the tail on male birds. Known as the cushion on the female.

SECONDARIES

The smaller wing feathers found inboard of the primaries, visible when the wing is folded.

SICKLES

The impressively long and curved tail feathers found on most male chickens.

SPLAYED LEGS

Also known as 'spraddled legs'. A condition affecting very young chicks which, in effect, dislocate their hip joints due to immature bones and a lack of muscle tone.

SPUR

The horny growth seen on the rear of most male chickens' legs, just above the foot. Spurs can sometimes appear on mature females.

SQUIRREL TAIL

A tail defect where the tail is angled forwards, towards the head, when the bird is viewed from the side. Regarded as a serious breed standards fault.

TRUE BANTAM

An exclusive group of about ten breeds for which there is no large fowl counterpart. Popular examples include the Dutch, Pekin and Sebright.

TYPE

The overall appearance of a chicken; its characteristic shape.

UNDERCOLOUR

The colour seen on a chicken when the main feathers are lifted to reveal the fluff at the base of the quills. This is another breed-specific characteristic in many cases.

WATTLES

Fleshy appendages on each side of the head, which hang down on either side of the beak. Usually larger on male birds and characteristic size varies from breed to breed.

WIND EGG

A tiny, usually yolk-less egg typically produced by a hen at the beginning or end of her laying cycle.

WRY TAIL

A defect where the tail is carried to one side or the other, rather than vertically, when the bird is viewed from in front or behind.

INDEX

FURTHER RESOURCES

BOOKS

Beebe, Terry. *Hatching and Rearing Your Birds*. Cudham, UK: Kelsey Publishing Ltd, 2010.

Beebe, Terry. *The Healthy Hen's Handbook*. Preston, UK: The Good Life Press, 2013.

Bland, David. *Practical Poultry Keeping*. Marlborough, UK: The Crowood Press Ltd, 1996.

Graham, Chris. *Avoid the Vet: How to Keep Your Birds Healthy and Happy*. Cudham, UK: Kelsey Publishing Ltd, 2007.

Graham, Chris. *Choosing and Keeping Chickens*. Cudham, UK: Hamlyn, 2006.

Graham, Chris. *Getting Started with Chickens*. Cudham, UK: Kelsey Publishing Ltd, 2009.

Graham, Chris. *Wisdom for Hen Keepers*. Newtown, CT: Taunton Press Inc, 2013.

Hams, Fred. *The Practical Guide to Keeping Chickens, Ducks, Geese and Turkeys*. Leicester, UK: Lorenz Books, 2012.

Roberts, Victoria. *British Poultry Standards*. Oxford, UK: Wiley-Blackwell, 2008.

Scrivener, David. *Popular Poultry Breeds*. Marlborough, UK: The Crowood Press Ltd, 2009.

Scrivener, David. *Poultry Breeds and Management: An Introductory Guide*. Marlborough, UK: The Crowood Press Ltd, 2008.

Thear, Katie. *Starting with Chickens*. Caernarfon, UK: Broad Leys Publishing Ltd, 1999.

MAGAZINES

Practical Poultry
Kelsey Media
Tel: 01959 541444
www.practicalpoultry.com

Fancy Fowl
Today Magazines
Tel: 01728 622030
www.todaymagazines.co.uk

ONLINE RESOURCES

Practical Poultry Forum
www.practicalpoultry.co.uk/forum

Rare Breeds Survival Trust
www.rbst.org.uk

The Poultry Club of Great Britain
www.poultryclub.org

The Rare Poultry Society
www.rarepoultrysociety.co.uk

FeatherSite
www.feathersite.com

IMAGE CREDITS